THE FIRST ORGANISM

IN SEARCH OF

AN ALTERNATIVE THEORY OF EVOLUTION

INNOCENT OUKO

THE FIRST ORGANISM

In Search of an Alternative Theory of Evolution

• • ● • •

INNOCENT OUKO

First print published 2023

Nairobi, Kenya

ISBN 978-9914-741-90-2

'No man ever steps in the same river twice, for it is
not the same river and he is not the same man.'

HERACLITUS

CONTENTS

i

PREFACE AND ACKNOWLEDGEMENTS

Good word was put in for Einstein when he was chosen to join the Prussian Academy of Science, following the rejection of his paper on quantization of light in 1905. This paper was dismissed by top physicists who then still did not believe it best to suppress what they believe was mistaken. This passage says of Einstein: 'That he may sometimes have gone too far in his speculations, as for example in his hypothesis of light quanta, should not weigh too heavily against him. For nobody can introduce, even into the most exact of the natural sciences, ideas which are really new, without sometimes taking a risk'. Risk. Indeed, I am taking a risk publishing this book. A risk firstly by introducing a theory or maybe at this stage, a hypothesis, which I believe explains Natural Selection from a perspective that to my knowledge, has hardly ever been explored. I am also taking a risk introducing this theory in simple terms rather than plundering the reader with the technical evidence I have accumulated in support of it. I take courage, nevertheless, knowing that from the start, most if not all theories are bound to fail one way or another. So why not write about it? As Karl Popper quotes in his book, *The Myth of the Framework*, 'One has always to take the risk of being mistaken, and also the less important risk of being misunderstood or misjudged'.

Put simply, the theory can be stated as: Using probability, physical existence is *sufficient* to explain evolution. One is therefore tasked with identifying this physical existence, and as this book explains, how to detect the presence of an organism. Probability would then complement this idea of the physical existence of an organism by giving any physical entity properties seen in biological organisms as we know them. However, in the interest of conveying the main message of the theory, a lot has been left out in this book. For instance, the handicap principle is much in keeping with my interpretation of mate selection, though it is more specific and granular. There are minor differences which if overlooked, still show how the principle explains what I emphasize with regard to size. As a result, it may look like I undersold other theories, but that does not mean I do not acknowledge their excellence. For instance, Mendelian Inheritance together with Darwinian Selection has been instrumental in paving the path for molecular biology. My profession, as a doctor and anatomist, requires that one is well versed with molecular human studies. Let it therefore not be taken that I have left out other important aspects in evolution. Epigenetics, for example, is the pivotal point where I started asking questions regarding the gene-centred view of evolution, which eventually led me to what we have now, a book.

I shall also add that it might appear that my theory is not scientific. A keen look nevertheless grants that it is empirical and by that definition, capable of being tested scientifically. For this reason, I shall note the statements which if disproved, might nullify my theory. It may need modification or it may be dropped as so fits any scientist. These statements, springing from my theory are:

1. There is no particle, and hence, no organism that naturally possesses *all* attributes similar to all attributes of another particle. Since we cannot tell if we have captured all attributes, we can limit it to the variables we know about particles such as mass, momentum, spin and spatiotemporal location.
2. There is no particle, and hence, no organism that naturally lacks the baseline cognitive property of *tending* to avoid annihilation.
3. There is no particle, and hence, no organism that is naturally incapable of forming mergers. A merger can be distinguished by the hierarchical concept where the role of one system serves another and vice versa. Thus, a merger constitutes a stable hierarchical relationship.
4. There is no particle, and hence, no organism that does not tend towards increasing in size or the physical equivalent of increasing in size.
5. There is no particle, and hence no basic organism, that is never in a constant state of motion.

These are the five postulates that embody the 4S's that I use to describe my theory. The first captures the role of Organismal Selection (OS) in demonstrating that it is a theory of diversity. The second one highlights the baseline definition and the operative means of identifying an organism. It also embodies the robust map of an organism and its universe, and hence upholds the second law of thermodynamics. The third highlights the fact that the same second law of thermodynamics can be interpreted probabilistically, and as a result, mergers will form. It underpins the basis of all known mergers, symbiosis and symbiogenesis as a primary engine in the formation of new organisms. In addition, it adds onto the basis of OS being a theory of complexity. The fourth displays the idea behind Natural Selection using reproduction as it is understood scientifically or duplication as a means seen in all organisms. Together with the fifth one, it elaborates on aspects such as Brownian motion and kinetic theory of matter. In short, the fourth and fifth shows how kinetic theory of matter and reproduction are potentially similar strategies of increasing in size. If any of them are revoked, then it is highly likely that my theory would be falsified or would need modification.

Another means of testing my theory would have to follow from the theory of algorithms, where we could consider organisms as compressed sets of binary codes, but I decided to leave it out of this book as it would constitute unnecessary technicalities for now. It would, however, form the basis for testing if particles indeed were organisms as a first approximation. Broadly, it assumes that each organism is a random algorithm. A change on any locus in the binary set would constitute mutation, but of unknown outcome. In such a state, mutations would have been catastrophic more than beneficial. The best strategy would then be to self-preserve rather than to change. This would predict what I have highlighted in this book regarding the kinetic theory of matter. This is just a single facet among several others which I would have to clarify in another publication. Nevertheless, the stem of the theory is still not lost for the reader.

Stuart Kauffman argues that Selection is not the only player in the game. Self-organization, which is spontaneous, explains how systems can exist in stable, ordered forms. If we imagine water in its three states – solid, liquid and gaseous, the solid form appears too ordered to appear 'alive', and the gaseous form appears too chaotic to appear 'life-bearing'. The sweet spot is the liquid state, which have been posited to exemplify biological systems. Such a state borders order (ice) and chaos (gas or water vapour). It is at the edge of chaos (liquid state). These are forms of order that are spontaneous so much that Kauffman considers these states as freely forming states, order for free.

We are all in a quest to understand some of the laws of self-organization, upon which Darwinian selection happens. The original theory by Darwin was that there must have been some common ancestor that descended with time through elimination (Natural Selection), gradually, to the rich diversity that we have presently. It, however, plagued Darwin how survival of the fittest, which acts through elimination of the unfit, should run contrary to the rich diversity witnessed in nature. Kauffman explains that self-organization answers in part, how complexity and as a result, diversity emerges. These then become the targets of selection. My theory takes a different approach, and arrives at the same results seen in all particular systems, either in combinations or solitary forms. It therefore argues that we are not at home in the universe, but that we strive to create one, by trying not to die in the first place. From the entry point of my theory, we witness how the difficulty of merging the two concepts of selection (as propounded by Darwin) and self-organization (as embraced by Kauffman) can be achieved. It is a possible step getting to understand how systems of order are tied to selection, which we have operationally argued to be annihilation.

To make these leaps in the field of evolution, I would have to do more than just publish my idea in a book. Experiments are needed, discussions, analyses and even formulating models in computer simulations. Thus far, it serves as more of a fertile conjecture than a stable fact or hypothesis. I would like to entertain the idea that if my theory is falsified, we would have made a significant step in getting to understand how we exist in our shared universe at least by offering perspectives previously not considered in science.

Finally, I would like to extend my heartfelt gratitude to all the authors whose works I have found useful in initiating, shaping and reshaping my ideas. Along this journey came Aditi Vakil, whose feedback I would always cherish in not only aligning the grammar but occasionally in challenging the logic of my thought. Adipo Sidang' for helping me get an editor, Dr. Tom Odhiambo and designer, Kariuki James Mwangi who have been instrumental in streamlining this final piece. My entire family, extended past my blood ties, for eagerly supporting and waiting for the final publication of this book. In particular, I am highly indebted to my mother Benta Oyoo, big brother Cohnrad Ouko and sisters Ivy Ouko and Dora Ouko who were the first audience who I introduced my ideas to, and who asked as many questions as they could. My small brother Al Christian and small sister, Yvette for tolerating my bursts of excitement through the early stages of formulating this theory. It is such an environment that challenged and encouraged me to pursue my line of thought. To close family and friends, Muntaz Mohammed, Edwina Kentagor, Limo Winnie and all my Rotaract and Rotary friends, Fadhila Richter, Thomas Amuti, Samantha Mukonjia, Joan Chumba, Khulud Noorani, Jimmy Njoroge, Jean Ojiro, Abednego Bingu, Trevor Ojiro, Khalid Mohammed, Lucy Syombua for being the first one to listen to the crazy implications of my theory, Prof. Moses Obimbo for being a superb mentor and many others whose names I cannot exhaust. As I finish, I would also like to thank you, dear reader, for considering this book. I hope it brings you the sense of wonder and excitement that it always brings me.

INTRODUCTION

As far as we know, evolution takes place only on our planet. We are made to believe that besides the other possible habitable planets discovered, ours is the only one where we are certain of life's existence. It may therefore be safe to claim that all else, that is, in our shared universe, is dead. That the universe has a knack for pushing for the boring. It just wants to kill all the excitement, and with excitement comes life.

For any high school student, the last week in school appears to have more days, and days more hours. My situation was no different. The year was 2011, the third and final term of the year, a week away from the December holidays. It was an unusually hot afternoon, and I was bored out of my wits with the year's Biology classes. The subject no longer excited me as it used to previously. For the entire year, we had been learning about ecology, which in high school largely entailed the binomial classification by Carolus Linnaeus. This paled in comparison to the fair lady, Physics, who had opened herself to me in ways that only lovers of the great subject know.

Nonetheless, for this particular afternoon Biology was the next lesson. Perfect. Unbeknownst to me, this would be the class that I would never forget: it would be my first formal introduction to the theory of evolution. My basketball coach and Biology teacher, Mr. Richard Ochieng', later went on to give an enchanting story of how Gregor Mendel discovered the laws of heredity of traits; how the humble, albeit inquisitive monk, diligently took account of the peas over successive generations, and calculated the ratio outcomes of the offspring at each generation to arrive at a fairly consistent relationship.[1] He did this, we are told, from mere interest and observation. In one fell swoop, Biology just became interesting again! For some brief moment, there was some excitement in the universe.

Fast forward and I am now at the University of Nairobi, and almost done with Medical School, but, for a good proportion of my classmates and friends, the topic of evolution has a galling taste. By this time, I had already developed what I still think is an alternative theory of evolution and was exploring, as Richard Feynman and Garrett Hardin insisted about matters theory or hypotheses, its falsifiable consequences. However, whenever the topic of evolution was brought up in a conversation, in my experience, the naturally abundant air thinned, tension rose and words got withheld. In other instances, the topic got ignored by short, sharp and final answers with little follow-up and no exploratory debate. It was often better never to bring it up in the first place. This was a sad eventuality since I loved and still do love the explorative adventure that evolution provokes. Furthermore, mentioning Richard Dawkins among close circles or even his affiliation with his seminal book, *The Selfish Gene*[2], raised what I always thought was unnecessary emotional conflict. I would later make peace with the inescapable conclusion that facts of the world do not have a care about our feelings. For instance, Dawkins and myself, were born in Kenya. That would not change. We also share an interest in evolution. I marvel at his wit and lucidity. Undoubtedly, he is a better writer, as will become clearer in this book. Also, our ideas about evolution differ. In the process, I hope not to make a rival of him or of any of his fans.

At this point, you might be thinking: who in his right mind would think that he has a theory to rival that of Natural Selection? Even more bizarre is: who in his sane mind thinks that he can rival the ideas of Charles Darwin, E. O. Wilson, Richard Dawkins or even Stephen Jay Gould? An undergrad in medical school? What I wish to share with the reader in this text is merely an alternative view of the currently held understanding of evolution. I respect the aforementioned giants and were it not for them and many others, I would never have understood evolution as precisely as they put it in their extensive writings. That said, in what I am about to explain, I have to admit that I am an amateur. I, therefore, merely state my arguments from that amateurish point of view. I gain courage, though, from great thinkers such as Pierre de Fermat, who considered himself an amateur mathematician, and Michael Faraday, who had little grounding in mathematics.[3] They nonetheless contributed profoundly to the fields of mathematics and physics, respectively. I also gain inspiration from James Lovelock, whom I will mention in various parts of this book, and who quotes Freeman Dyson when he states that the ethics of science is based on a fundamental open-mindedness.[4] To this statement, Lovelock adds that creative science requires a sense of wonder and humour. I have laughed and marvelled at how my theory has reshaped my thinking and I hope you too can laugh it out, at the least and marvel at it, at best. Thus, while every scientist hopes to achieve such levels of acclaim as the greats I have mentioned, I merely intend to stir thought among evolutionists and lovers of knowledge.

I have to insist that the reader understand that what they are about to discover is different from what might be perceived as common sense. This much I can guarantee almost in the same way Nicolaus Copernicus insisted that the Earth was not the centre of the universe. The central aspect that Copernicus introduced was that he made people start thinking that the paradigm might be different, by insisting that the laws that govern celestial bodies might be at odds with common sense. So many of the laws that have evolved ever since the Copernican revolution have had their tumultuous journeys and even barren phases of breaking through the prevailing scientific ranks. Much of what I will try to convey here will appear to be at odds with some current paradigms of evolution. However, the unification of disciplines such as physics and biology that this theory attempts to achieve is a good enough reason to proceed with the idea. These two disciplines are essential in the field of complexity. As such, my theory in part touches on complexity.

I shall begin by establishing some ideas that will hopefully become clearer by the time you come to the end of this book. First and foremost, the gene. Arguably, the fundamental unit of *heredity* is the gene. In my theory, I propound that the fundamental unit of *diversity* is the organism. The theory that I hope to introduce you to is, therefore, in part, a theory of diversity. I hope to guide you into seeing that my theory explains, albeit differently, currently-held theories of evolution. By focusing on the organism, we will appreciate a different picture of evolution, a *fractal* picture. The fractal geometry of an image or object is a property of self-similarity that appears across smaller and smaller scales of the image or object – in short, the large and smaller scales of the object appear self-similar. I intend to explain how evolution is self-similar at all scales, structurally and behaviourally. The theory also explains the emergence of consciousness and life, as we vaguely know it as well as the various incidence rates of cancer among rats, humans, elephants and even the blue whale. It will also expound on the phenomenon of growth and aging.

I specifically have to highlight one facet of the theory that I am endeared to – reproduction. While Natural Selection explains that an organism has to have reproductive capability, it hardly explains *why* they have this ability. My theory explains why reproduction happens, why it happens at the age it does, and further explains menopause as a unique feature among human beings. In fact, it attempts to show why the term 'reproduction' is a misnomer. Here is a little hint: it is related to size. Size will, thereby, be the single aspect of my theory that contains a vast predictive value. In fact, after reading Geoffrey West's book, *Scale*, I immediately dove into writing this book as West presented sufficient evidence that I needed in support of my theory, even though he never considered the perspective of organisms that I had in mind.[5] To echo West's emphasis, at this point, I can only think that what I have is a *coarse-grained theory*, but with potential for a lot of experiments.

At its inception, the theory I propound was birthed from mathematics, but not in a way that inspires nightmares, as the subject is generally taken to. It will be easy for any reader with a basic foundation of Mathematics to understand. Most importantly, however, is the relationship between the organism and its universe. I repeat, the theory of diversity cannot be appreciated without emphasizing this consistent relationship, which I will also elaborate throughout the book. It is against this robust background relationship that diversity buds. Appreciation of this relationship starts by going against the hailed scientific thinking, which has always started by identifying what any substance is made up of. It is what some might call analytical thinking. This approach has led to identification of the genome, the double helix, the Mendelian laws of heredity and exceptions that bypass such a mode of inheritance such as the various epigenetic inheritance systems. It has led to the identification of diseases and preventive and curative approaches in defeating them. This type of thinking has achieved a lot scientifically. However, it has certain shortcomings – for instance, it encourages a predominantly *linear* perspective to the point of various scholars calling it reductionist. Organisms, on the other hand, display *non-linear* features. Also, in the most rigorous of ways, scientific thinking stresses the substance, but as Brian Goodwin reminds us, composition is not sufficient to determine the form. My theory has a different focus, central to which is the relationship between the organism and its universe. The key word is *relationship*. It does not stress so much the substance of the organism but the relationship between the organism and its universe. Sadly, relationships, unlike substances, cannot be measured nor weighed in the sense that science has always done. To appreciate this theory, therefore, the reader should occasionally take a step back from the scientific recesses of measuring and weighing, and remember the relationship.

Fortunately, while relationships cannot be measured or weighed, they can be mapped. The map that I will draw in your minds is so simple, it me baffles to date. For instance, the map allows both the top-down approach as seen in systems thinking and the bottom-up approach as evident in the scientific method. In fact, it uses some bottom-up elements to validate the necessity of considering this theory as scientific. The keyword here is *consider*. In addition, what I hope you will come to acknowledge from the map is that although the substance can change, the relationship remains constant. Thus, using the relationship as a constant, the diversity, which I fundamentally attribute to the organism is easily appreciated. The result is the emergence of resilience – diversity will be seen to achieve resilience. A central take-home would be that organisms evolve for resilience. Resilience will take many forms such as the much studied dynamic disequilibrium states. For instance, the Earth displays various resilient properties, the result of which is a fairly stable earth temperature and invariant proportions of gases in the atmosphere for long periods of time. My theory thus regards the earth as an organism, much the same way as Lovelock sees it.

But first, we must ask ourselves, why a new theory?

CHAPTER ONE

Why a New Theory?

Long live the idols, may they never be
*your rivals – **Let Nas Down, J. Cole***

In his book, *The Language Instinct,* Steven Pinker speaks about the anchoring limits of current views regarding evolution.[1] If an entity is a product of nature and nurture, in the colloquial sense, why then should we anchor on the gene alone? If these organisms get socialized, amounting to nurture, why should we anchor on the group alone? Why pick sides and decide to either go with a gene-centred view or a group-selection view? Why not take an option that considers both? I concur with Steven Pinker when he states that preferring one side blurs or even casts a blind eye on the other. He equates it to focusing on the Cheshire cat, without the grin (which is its identity) or the grin without the cat (which Alice found absurd).

The gene-centred view focuses on the inheritance of traits from the parent to the offspring, and from generation to generation. If I have black skin, this trait is likely to be passed on to my offspring. These and other traits are captured in a transmissible code known as the gene. Therefore, the gene-centred view elaborates that the fundamental unit of heredity is the gene. This view is conventionally accepted by many evolutionists and biologists.[2] But that has been the central focus for a long time. What of variation? Since the focus has largely been on the gene, the currently held explanations have attributed organismal variation to the errors that arise from the process of replication of the molecule where the gene is archived, the deoxyribonucleic acid (DNA). For the gene to be inherited, the code has to be 'remade' before it is passed down to the offspring. The 'remaking' process is often marred with errors that have been used to explain the causes of variation. The other cause for variation is related to gene mutation – the functional change of the gene. Some have even attributed high degrees of variation to the process of recombination through sexual reproduction, which brings together two people's dif-

Alice and the Chesire Cat. The Cat, known for its grin, is identical to the genome that in Biology, distinguishes one organism from another. However, exclusive focus on the genome and the gene is similar to focusing on the grin and not the grin in the context of the Cat itself.

ferent genetic make-ups, yielding an offspring that is uniquely different from its parents.[3] This last subtle process holds the key to understanding the theory that I intend to propagate – that an emergent being was formed from two compounds fusing. The kind of questions we should ask ourselves from this process should be: why did sexual reproduction evolve in the first place? Why did such a relationship develop, where one sex depends on the other? By the end of the book, I hope to have fashioned another reason besides the gene-centred idea of variation. I also hope to convince you that size plays a role in the evolution of sexual reproduction. Since my theory is a theory of diversity, it propounds that we do not necessarily have to yield to the imperfect process of replication for there to be variation or diversity.

Substantive evidence has confirmed that coded information is passed in parts from one generation to another through genes. For evolutionists, a gene merits relevance only after it lasts long enough to produce copies of itself, and then get to compete through Natural Selection. Hence, while the gene-centred view focuses on the *fundamental unit of heredity*, I propound that the *fundamental unit of diversity* rests with the organism. I call this theory the theory of Organismal Selection (OS). At this point, I hope the reader has at least acknowledged or intuited a difference between the two theories.

There are more reasons, however, that call for a modification of the gene-centred view or the formulation of an alternative theory altogether. For instance, according to the gene-centred view of evolution, the coded information is captured by the DNA molecule. Therefore, the limits of this theory are bound by the timeline when the world allowed the existence of such molecules, a world known as the ribonucleic acid world, or in short, the RNA world. But the gene is not the DNA or the RNA. It is a part of it. In particular, it is a code. In our world, as we know it, this code is archived in the DNA/RNA. Thus, Richard Dawkins was very safe by insisting on using the word 'gene' in this sense, so that his arguments remain valid throughout his famous book, *The Selfish Gene*. That begs the question – before the RNA world, did the code exist? Organismal Selection (OS) hopes to offer a possible answer.

What about group selection? Unlike gene selection, group selection is the theory that Natural Selection will favour traits in an individual that makes the individual's group survive. It has always bothered me why it should be the superior candidate theory in evolution. There is little doubt that this theory explains several aspects appreciated in nature.[4] However, traced back to the very beginning of life, it does not explain how the first organism evolved. Or even the second. The first organism is not in any sense a group. The first organism should, however, be subjected to the laws of evolution. Therefore, Organismal Selection (abbreviated as OS from here henceforth) *may* be better suited to explain the state of the first organism or organisms before any 'group' was formed. It also begs the question: does the first organism constitute the original gene pool upon which Natural Selection acts, and if so, is this a 'long enough' time before Natural Selection acts? And if Natural Selection acts on what is available in nature, why did the first organism exist as it did? Why did it take the form that it had at the very beginning? We do not know the form that the first organism had. However, this should not stop us from questioning why it existed in whichever form that it did, a form that cannot be explained by Natural Selection. Even after mapping all the genes in an organism, it is still not sufficient to describe its definitive form.

The idea of the first organism therefore not only introduces unexplored aspects in group-selection theory, but also elucidates a paradox for the gene-centred view concerning its definition of 'long time'. The first organism could have appeared and disappeared, and succeeded by others which could have also appeared and disappeared in the span of time that each existed. The idea of whether it was a long time or a short time becomes secondary to the fact that they existed in the first place. In the gene pool, a long time is millions and even billions of years, but for the first organism, this might not have been the case. As a rule, first, the entity has to exist whether it is a gene, an individual or a group before any of the theories can be discussed. But before we dive deeper, let us revisit the tenets of Natural Selection.

Even after reading Dawkins' *The Selfish Gene* and *The Blind Watchmaker*[5], I still did not have a clear idea of what Natural Selection was. Steven Pinker is the first person who, in my reading, described Natural Selection succinctly. Natural Selection is the merging of two independent ideas: that of reproductive capability and, birth and death rates. An organism bears the ability to reproduce and pass down coded traits to the offspring. These organisms, when they reach a certain number, die at different times, constituting what might cumulatively be acknowledged as the death rate. The death rates will be more for those organisms that have unfavourable traits and birth rates will be maintained or increase for those organisms that have favourable traits. These are inescapable facts that we have to contend with. It is one of the reasons Natural Selection is so potent.

But the current understanding of Natural Selection is from the perspective of the gene pool, where genes emerge and compete with others for a period long enough before Natural Selection acts. By focusing on the long-term existence of a gene, the emphasis of a gene-centred view of Natural Selection is more of a game of competition. What of cooperation? Dawkins shows that despite the title of his book, *The Selfish Gene*, cooperation emerges even among the strictly selfish genes. So cooperation and competition tango in life as we understand it. With reference to the first organism, however, it forms a quandary.

If we had a first organism, which we clearly did, then it does not explain why it should compete. Compete with whom? So competition is out of the question. What of cooperation? Cooperate with whom? Cooperation is also out of the question. The idea of the first organism, therefore, pokes holes at the central forces that shape the predominating theories of evolution. A different explanation is called for – an explanation that precedes the emergence of the first organism that might be understood currently as the common initial ancestor. For the first organism to be subject to the same laws as its subsequent and present organisms, we have to develop a theory that predates organisms as we have always understood them.

From a gene-centred view, individual organisms play a minor role in the evolutionary spectrum. If life emerged on Earth over 3 billion years ago, the living organism such as you or me, is virtually non-existent in that time scale. This is one of the central reasons why the gene-centred theory focuses on the long-term effects. Genes last longer than organisms. Such a broad lens of evolution does not capture the details that happen over the brief moments when organisms exist, or the point of noting when a novel gene arises. Therefore, the arrival of the fittest gene has not been a well-established concept (although Andreas Wagner[6] has done a substantial job trying to solve this problem). Organismal Selection (OS) takes a tangent and focuses on these neglected pieces. It focuses on the solitary pixel that is neglected in the interest of the big picture and argues that the organisms are pivotal to the existence of the gene. This might sound intuitive and not worth the bother, but OS extends past this intuitive feeling to not only describe the emergence of the gene, but offer a different perspective on what an organism is. In this sense, from one organism to the next, we get to understand the ruggedness of evolution, the fractal nature of evolution.

Also, the question about the emergence of life and consciousness remains a mystery. While these are the commonplace discussions about the shortcomings of the neo-Darwinian gene-centred theory, there are other puzzles that I have also not found solutions for. They are as follows: one of the principles of Natural Selection implies that at the very beginning, the organism was geared for reproduction. This is an assumption that we yield to because currently, all organisms that we have explored exhibit reproductive behaviour and capability. This assertion does not match the evidence that we draw from evolution itself: that the most optimal strategy succeeds the other less favourable ones. If we are to go back in time, it should make sense that reproduction be a strategy that competed with other strategies, and eventually, proved the more favourable one at the time. Therefore, in a thought experiment, the currently held idea of reproduction as an essential feature in evolution breaks down the moment we imagine the presence of other strategies competing with the reproductive strategy. I later argue that *reproduction* is a misnomer, and introduce the process of *production*. Consequently, I must also acknowledge the additional assumption I will be making - that the first organism had other strategies or other strategies besides reproduction. So, reproduction was either present from the emergence of the very first biological organism and proved to be superior, or it emerged later. I believe that this is a valid assumption, since we do not know how it began. We only know that life must have had a common ancestor. The argument that I make would invalidate the need to consider other strategies besides just one: *production*.

In addition, since Natural Selection is a statistical theory, where the death of the unfavourable organisms is succeeded by the survival of the favourable organisms, it also does not apply to the first organism. If the first organism dies without reproducing, it does not get replaced by another organism. It is gone. In such a scenario, there is no *rate* in the sense of Natural Selection, when there is only one organism – it exists or dies. In a universe such as ours (whatever that is), and as history dictates, this is a plausible picture, where an organism dies before reproducing. Even then, if it existed, the first organism should have been 'reproduced' by another that preceded it, creating an infinite regress that does not explain how the first ancestor came into being. Thus, the idea of death rates and 'reproduction' invalidates the theory of Natural Selection with reference to the first organism. Ironically, this elusive common ancestor is the common link to all the hypotheses explaining Natural Selection. But the very first organism should also be subject to the laws of evolution. The concept of the first organism, therefore, sheds light on some cracks in the edifice of the majestic theory of Natural Selection. I hope to convince you that the alternative theory, Organismal Selection, does not break at this point. However, we are not done with dissecting Natural Selection just yet.

Darwin borrowed the idea of Natural Selection from artificial selection.[7] By observing how farmers cared for their plants, selecting the good ones from the not-so-good ones, there was the overall survival of the plants with favourable traits. What I have never agreed with was the need to differentiate between *natural* and *artificial* selection. It assumes that human actions are not natural. This is unwarranted, since the selection process is significantly a matter of chance, as Gregor Mendel, the founding father of genetics, proved. If I decide to build a concrete house, how different is it from a beaver building a dam or a bower-bird decorating its home with pebbles? I have never seen the need for having this distinction, even though it helps to distinguish the effects of our actions. As the beaver builds the dam and affects the flow of water downstream for the local villagers, so does the local man build a house using the logs of wood the beaver would have used to build the dam. But the need for the distinction has taken root so much as to have man distinguish himself from other creatures and from non-living things.

This need for telling man apart from other material could be traced back to the days of Aristotle, or perhaps earlier, to his teacher. From the teachings of Plato, man was seen as a superior being and an animal as a lesser being, which could be described as just that, 'animate'. Aristotle then gave man a conscious essence, a consciousness, which was matched with his capability for rational thought.[8] Animals, he conceded, were not capable of rational thought. Thus, the need for division between man and the rest continued. Presently, theories of evolution distinguish themselves from theories of other sciences such as Physics. In fact, cultural evolution, which has shaped the world in such a short span of time, is usually distinguished from biological evolution by the adjectives 'cultural' and 'biological'. Firstly, as distasteful as it might sound to the traditionalists, Physics is an environmental science. The environment comprises humans and their surroundings. Therefore, the theories of evolution ought to agree with the theories of Physics. The dichotomy between living creatures and non-living creatures has been a major barrier between the two fields. When Stephen Hawking talked about the theory of everything, he actually meant everything[9], not just non-living material as we know it. The theory that I wish to propound seeks to erase the dichotomy between the animate and inanimate, and once more show that the living and non-living are particles of nature. The distinction between natural and artificial processes should then be seen as a smokescreen, albeit a useful one, that shows our effects on the environment. I will, therefore, unlike a good proportion of previous works on evolution, use evidence from fields supposedly distant to evolution (biological or cultural). I shall then go back to the vast evidence that the field of evolution wields, and leave it to the rest of the world to falsify my theory.

Another problem that has been raised with regard to the theory of Natural Selection is the property of passivity.[10] Natural Selection, according to some scholars, is rather passive. I hope to paint a different picture, of how active the entire process of evolution is and how the active bit can create the illusion of passivity. This can be achieved by stretching the theory of OS a little by introducing the reader to Per Bak's concept of self-organized criticality.[11] This concept shows how the organism is very much involved in the production process. More about this will be discussed in the latter chapters of this book.

The other problem is rather a personal one. I find that Natural Selection acts like a fail-safe, or a blanket statement that everyone can always resort to in explaining away the traits that they encounter in new and old species alike. Even more, the gene-centred view has taken root so much that any form of variation or even heredity tends to be explained away by genes. It might appear that I am outright against the prevailing theories of evolution, but I only try to show how they are relatively good approximations of what we observe on Earth. To appreciate this, I shall develop OS and compare and contrast it with the current theories. So, without further ado, let's jump right into it.

CHAPTER TWO

Organismal Selection

*Hoping my pen lays my name down, until that day's found, this is my playground – **Playground, J. Cole.***

First things first: a statement concerning our ignorance. In this Chapter, I will be using probability in the simplest way that I can think of. And probability is a statement of ignorance – an admission of our ignorance. The likelihood that a coin toss can either land on heads or tails does not predict that, on the first toss, it will land on heads and on the second, tails. Any of those sides can happen when a coin is tossed. Probability statements are close-to-accurate statements of our ignorance. Nevertheless, the most tested and robust theory yet, the theory of quantum mechanics, thrives despite its reliance on probability. It embodies the spirit of science – curiosity and modesty. Secondly, the testability of my theory to its fringe extents is a feat I can never fully explore. That again is an admission of my ignorance. Thirdly, evolution. John Maynard Smith once said that everybody claims to know what evolution is. I do not remember where I read this but I do remember that it resonated with me. Even after reading Dawkins works, I still did not feel like I understood it. I had to read several books by other authors before I understood it (I think) very well. But, ironically, in light of all this darkness about evolution, I intend to give it a definition that I believe is much better, one that gives enough wiggle room to incorporate materials such as cells, atoms and quarks.

On the 30th December, 2019, I entered the search words 'evolution definition' and this is what Google delivered:

Noun – 1. The process by which different kinds of living organism are believed to have developed from earlier forms during the history of the Earth. Synonyms: Darwinism, natural selection;

1. The gradual development of something;
2. The giving off of a gaseous product, or of heat;
3. The pattern of movements or manoeuvres;
4. The extraction of a root from a given quantity.

Clearly, evolution has different definitions. Our main concern is: what is it in terms of biology?

So far, the definitions have worked to distinguish what evolution is in biology from the other sciences. I reckon that there is no need to have such a marked difference. Thus, the definition I work with would appear to be applicable in biology and other fields such as physics and chemistry. The definition I run with is that evolution is *the ability to avoid annihilation*. Simple. The story of evolution then becomes the story of how organisms were and are able to avoid annihilation. If you are into genes, then it is the story of how genes were and are able to avoid annihilation. This rather morbid definition has far-reaching consequences besides the fact that it captures the previous definitions. Taking a gene-centred approach, it is the ability of a gene to avoid annihilation, hence its *self-ishness*. From the species or group level, it is the tendency of the species or group to avoid annihilation hence the propensity for reproduction. To capture the words of Lynn Margulis, 'Natural Selection eliminates and maybe maintains, but it doesn't create'.[1] Thus, if Natural Selection is concerned with elimination, then annihilation is a central concept in evolution. The theory of OS takes up this definition. That is, evolution is the ability to avoid annihilation. It is, therefore, ironic how the morbid definition helps us understand and appreciate an alternative perspective about life.

As the ability to avoid annihilation, evolution bequeaths every entity that can evolve a probability of evolving. Thus, the consequence of this definition can be extended to other realms that distinguish biological evolution from other evolutions. When Stephen Hawking and Leonard Mlodinow talk of how the universe has evolved, they build the plot for introducing the success story of how quantum mechanics has told a very different but exceedingly accurate story about the universe.[2] In their book, their focus is the universe. Biology narrows the scope to study objects of life. It deals with the objects that are a part of the universe. Physics concerns itself with particles and energy. It also deals with objects of the universe. Having a definition of evolution that excludes one from the other, that is, the evolution of organisms from evolution of the universe, while convenient in academia, often stops one from appreciating such similarities. Taking evolution as "the ability to avoid annihilation" intends to form a fairly stable bridge between the two fields and seeks to nurture a healthy relationship between the two disciplines. Let us explore, superficially, the possibilities of such a relationship.

Physics attained a new level of growth when Isaac Newton and Gottfried Leibniz independently discovered calculus.[3] Experiments then began to have a quantitative aspect. Physics and Mathematics have since been inseparable. By introducing such a definition of evolution, I hope if not to create a similar bond, to emphasize its relevance, and incline Biology towards a relationship similar to that of Physics and Mathematics. What this would mean is that Biology and Mathematics would also be seamlessly fused. By sharing these ambitions, I do not mean that there is no relationship, but that the boundaries of academia have insisted that each field remains different. In fact, there are fields such as Mathematical Biology that have advanced evolution into unexpected spheres such as Game Theory.[4] I identify with Judith Rich Harris when I take this position, of not being bound by the territorial boundaries of academia.[5] By building on the argument for OS, I will be shifting from one field to another. In fact, it should lead to a different understanding of what an organism can be. But what does Biology tell us about organisms?

Biology teaches us that an organism can be unicellular or multicellular. The cell is therefore the basic structural and functional unit of life. It is this life that we do not know how to define. If life were to evolve,

Biology states that evolution would have occurred if favourable traits were inherited by the offspring and its subsequent progeny. Organismal Selection views it differently. An organism, which should be subject to the laws of evolution, is bound by the morbid definition that we accorded it – the ability to avoid annihilation. It displays this ability by existing for a particular period before it ceases to exist as we know it. Broken down further, this organism comprises particles. Each of these particles forms a nexus that constitutes what we ultimately understand as an organism. But each of the particles has its own abilities to avoid annihilation. Therefore, when the organism dies, the features we accord it, what Biologists call life, also cease, even though the particles remain. Organismal Selection recognises that the living organism is a composite of the various particles that constitute it, each with its own ability to avoid annihilation. The particle and the living organism have their distinct abilities in avoiding annihilation.

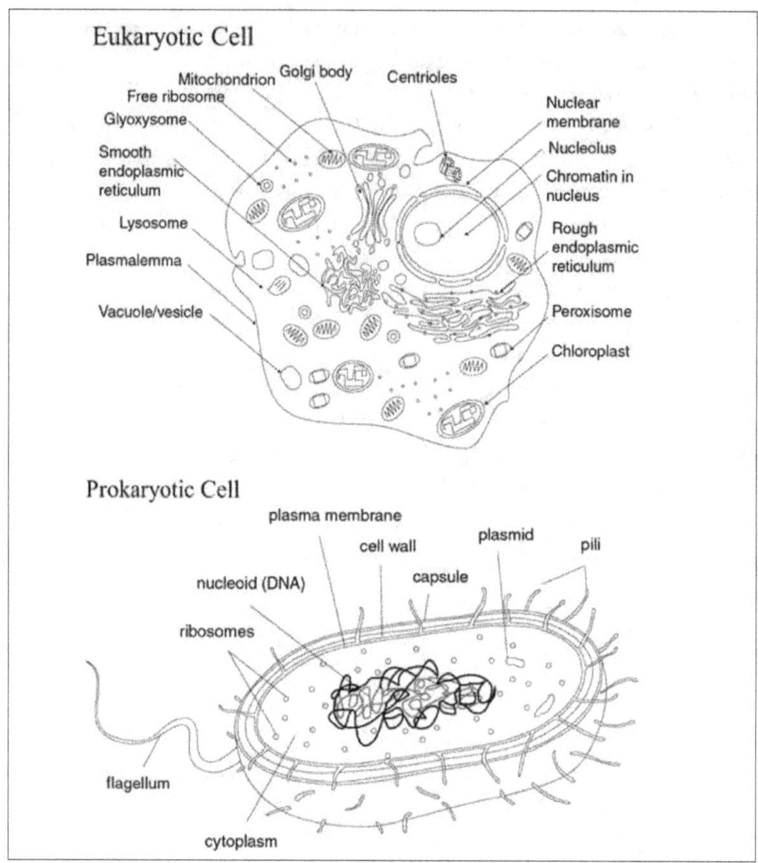

Eukaryote and prokaryote. The former defines organisms with membrane-bound structures inside it, while the structures inside the later lack membranes.

Particles and their inherent ability to avoid annihilation

When particles aggregate or chemically combine to form a mixture or a compound, the resultant combination of these particles forms a singular entity, what I call an organism. I call it so since it now has a renewed or altered ability to avoid annihilation. From our definition, what does this say about what an organism is? The simplest organism is a *particle*. It contains the two properties that constitute an organism as we know it – existence and ability to avoid annihilation. Absurd? Maybe at first glance, but the same properties still hold for both particles and living organisms as we understand them. My interest is in the organism – whether you want to call it *living* or *non-living* is up to you. I hope to take you from absurdity to possibility. This definition of an organism is only in part, since it covers what I call the *structural component* of the organism.

The other bit is the *extra-structural* component. This component is rather simple. It is better appreciated as a function of probability. The extra-structural component of an organism (particle) is the *probability of annihilating the organism*. Let's say the probability of annihilating particle **W** is 1/3. Consequently, the probability of not annihilating it is 1-1/3, which is 2/3. This probability function accords a particle with the property that subjects it to evolution: the *ability to avoid annihilation*. As a consequence of this extra-structural property, every particle has the inherent ability to avoid annihilation.

It is important to stress that this probability is not equal to zero or 1. By converting a particle's ability to get annihilated into a probability function we accord it a *tendency* to avoid annihilation, which, again, lies between these two limits, 0 and 1. The simplest organism is therefore *a particle and its tendency to avoid annihilation*. This takes into consideration the structural (particle) and its extra-structural (tendency to avoid annihilation) components. In this simple way, the first organism is subject to the laws of evolution. It does not need to compete with any other organism, nor cooperate. It just has to exist, like the common ancestor.

There is another bit that has not been explained. When introducing yourself to others, you would start with the words *'my name'* and not *'our names'*. Alternatively, you would start by using the pronoun *'I'* – I am a student. Emphasis on the singular identity (my/I) rather than the plural version (our) stresses the relevance of the extra-structural component. How it works, therefore, is that as a *particle* the "tendency to avoid annihilation" is a function of the *particle*. But as a *compound*, the "tendency to avoid annihilation" is a function of the *compound*. Therefore, while particles combine, the tendency changes. Let's say the probability of annihilation of particle **A** is 1/3. That of particle **B** is 1/5. When the two particles combine the probability of compound **AB** being annihilated is equal to the probability of annihilating **A** *and* the probability of annihilating **B**. That is, 1/15 (1/3×1/5). In probability the function 'and' implies multiplication (1/3 × 1/5). The result of this is that when the two particles combine the tendency to avoid annihilation is 14/15 (1-1/15). It can only acquire this new tendency if the resultant combination of the two particles forms an emergent entity, in this case **AB** or **BA** or **Sodium Chloride** or **Innocent** or whatever you may call it. From probability, it shows that this new entity, the new tendency, is different from the original one from each particle. This emergent component, in OS, accounts for the singular identity (*my name*), since the emergent tendency to avoid annihilation (14/15) remains singular. If this entity, such as the one who is currently reading this statement, is conversant with the rules of English, it can use a singular pronoun 'I' or 'My' instead of 'We' or 'Our'. Conjugation of the two particles, therefore, results in an emergent organism. It qualifies to be an organism structurally but even more importantly, extra-structurally!

In order to accept this formality we have to acknowledge two important caveats. Firstly, that the original constituents, **A** and **B**, maintain their original probabilities the moment they merge. However, this is a formal angle but formality does not perfectly correlate with reality. The merger would have to be *perfect* in every way for the particles to maintain their original probabilities. As we shall later see, organisms satisfice rather than optimise; hence a perfect merger is not possible. Secondly, the new probability of annihilation reduces only when we consider annihilating *all* the organisms – that is, **A**, **B** and **AB** (or even **BA**, with regard to the ordering of the particles). This assumption is important since **AB** can easily be destroyed by simply chipping a portion off it. The new probability of annihilation has to reduce only when we consider destroying *all* the constituent and emergent organisms.

If we assume that we do not know the probability of annihilation of a particle, we can work with a derivative of this example. We can assume that two particles have the same probability of annihilation. Particle **X** and **Y** can have this probability, and we can even give it an arbitrary probability, say ¼. When the two particles combine, the probability of annihilation shifts to 1/16 (¼ × ¼). It can only shift if the result is a new organism, say **XY** or **YX** or **Oxygen Molecule** or whatever you would want to call it. It is only by acknowledging it as a *new organism* that it acquires this new tendency to avoid annihilation – that is, 17/16 (1-1/16).

I have to insist that by new tendency to avoid annihilation I mean *complete* annihilation of the new organism *and* the original constituent ones. It is only then that the emergent probability can be appreciated. We can, therefore, create a general rule – the more the particles combined, the less the probability of annihilation and the greater the tendency to avoid annihilation. Hence, the probability is inversely related to the number of particles.

We thus arrive at the general formula:

$$P_a = \frac{1}{N}$$

Where P_a is the probability of annihilation, and **N** is the number of particles. This formula is, however, a tad flawed because if one takes the example of a single particle it would mean that the probability of annihilation is 1, which is absurd and empirically unsupported. The formula is a *general one* that goes only to show that every particle has its own probability of annihilation, which I must insist, is never constant. I will later explain why this probability is not constant. At this point the formula is only abridged because in reality the probability of annihilation is not a constant. It is not a *fixed number*, but a *fixed property* that is nevertheless present due to the existence of the particle.

This property further gives room for the definition of annihilation. Regarding existence, annihilation can simply be defined as *the cessation of existence*. For a particle, the simplest organism, annihilation is the cessation of that particle's existence. This is the simplest definition of annihilation I can conceive. I will later explain how this definition can be expanded to include living organisms as we understand them. For now, cessation of existence suffices. It takes us back to how we defined the simplest organism – a particle and its tendency to avoid annihilation. Defining an organism in this gloomy way speaks of an eventuality – that organisms will eventually be annihilated. This final destination for every organism shows that the probability of annihilation cannot be a constant. It has to shift towards the upper limit of 1 – from *probable* to *certain* annihilation. Thus, the mere existence of a particle implies it contains a probability of annihilation and some tendency to avoid it.

Why should we call it a tendency? Imagine this scenario – you have a particle with a probability annihilation of 1/7. This probability increases by 1/7 every day, so that the particle has one week (7 days) to live or exist. This gives it a probability of avoiding annihilation of 6/7 on the first day, which turns to 0 at the end of the week. If the particle were passive, it would simply just exist and wait for its doom. However, it does not do that. As I shall expound on later in the book, particles are constantly in motion, seeking mergers that eventually postpone their annihilation. Now, assume that it gets another particle on the first day, whose probability of annihilation is also 1/7. When they merge, the resultant probability is 1/49 (1/7×1/7). This delays both their annihilation by 42 days (49-7). The *tendency* to avoid annihilation gives the particle the same property as any organism as defined in Biology, an organismal property.

There is another subtlety that I have to emphasize. Let us say we have two organisms, **A** and **B** that merge to form either **AB**, **BA**, **Heisenberg** or whatever name you wish to give the emergent organism. Such a system comprises three organisms – **A**, **B** and the emergent organism. Initially, the original constituents were progressing towards certain annihilation but formed a merger. The resultant organism therefore will also have the same predicament, progression towards certain annihilation if it also does not form a merger, or find other ways of avoiding annihilation (as I shall elaborate in the latter chapters). So, the emergent organism can very easily get annihilated if it does not up its ante. As for the constituent organisms, **A** and **B**, they delay their annihilation since they already formed a merger. Henceforth, while it might be easy to annihilate the emergent organism, this process delays the eventual annihilation of **A** and **B**. Each organism whether constituent or emergent has to find a means of avoiding annihilation, of which a merger is a viable option. However, the argument that OS offers is that *complete* annihilation of the three organisms can be appreciated using the calculus of probability.

Now, the annihilation *probability* of a particle shifting towards *certainty* introduces the first facet of diversity. That is, a particle will have its *own* property of annihilation always shifting from probable to certain annihilation. A particle will have its *own* temporal shifts from existence to cessation of existence. The probability will therefore be unique in how the particle first appears and then later disappears. Organismal Selection, as a theory of diversity, states that the fundamental unit of diversity is the organism. Using this probability function we have, therefore, shown how every particle is diverse. There are no two alike.

The general formula we introduced earlier (i.e. $P_a = 1/N$) is important *only* in showing that the more the particles merge, combine or fuse, the lesser the overall probability of annihilation becomes and the greater the overall tendency to avoid annihilation. The hydrogen atom and my next-door neighbour share this common property – a tendency to avoid annihilation. At the very least, it allows one to consider that a particle is an organism. My neighbour exists and that particular hydrogen atom exists. They have both been able to avoid annihilation thus far. Surely, they must have some properties that have enabled them achieve this feat. Evolution should be able to tell this story.

The cautionary take-home message from all this is that we do not need to know the particle's probability of annihilation. In truth, even from experimental tests, we do not know what this precise probability is. But the formula shows that we do not need to know – we only need to know how it works. The simplest way of avoiding annihilation is by combining, fusing, mixing or forming a compound with another particle. We have Chemistry, Physics and Biology that show how combining, fusing, mixing or forming a compound is a ubiquitous trait.

Let us try fine-tuning the initial general formula to a more precise one. By combining, fusing, mixing or forming a compound with another particle the probability of annihilation reduces. This means that the tendency to avoid annihilation (T_a) increases. The tendency to avoid annihilation then becomes 1 – (the product probability of annihilation of the particles) as shown in the formula:

$$T_a = 1 - P_a$$

Where P_a is the probability of annihilation. This second formula is very simple and, in principle, more precise than the previous one. It acknowledges that if the probability of annihilation has shifted from a fraction to 1, then the tendency to avoid annihilation is 0 – the organism no longer exists. Should we, perchance, know the probability of annihilation of the particle, then the general principles of probability remain unchanged. The simplicity of this rule is evident even in elementary examples. If I am to annihilate two magnets, I would need energy to first separate them, then destroy them individually. By the two magnets being attracted to each other they demand that the force of annihilation becomes much more for the eventual annihilation of the composite organisms (the two separate magnets). You will need more energy to separate two pebbles that are close together before crushing each separately. This process of destruction, that is, separation, exposes you to more small parts of an organism (in principle, another organism) that need subsequent destruction. Combining, fusing, mixing or forming a compound delays annihilation of the constituent organism, enabling them to temporarily avoid its previous course towards annihilation.

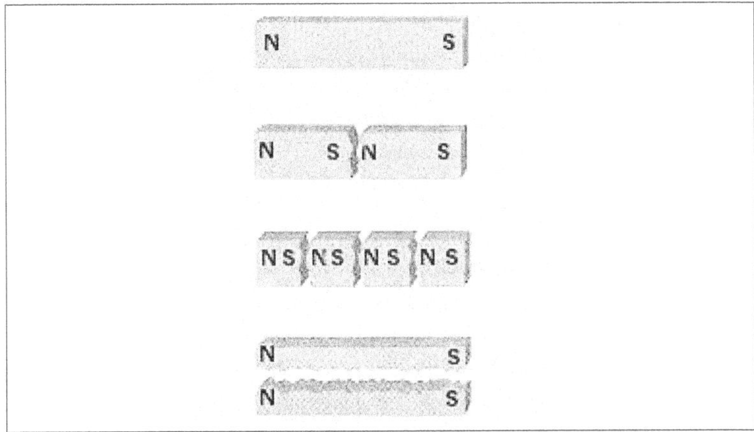

Magnets split across different axes. More energy is needed to separate magnets into different smaller magnets. Interestingly, magnets preserve their polarity.

Organismal Selection, thus, defines the simplest organism as a particle and its tendency to avoid annihilation. The implications will be tested in subsequent pages and chapters of this book. At this juncture, we have only begun to establish OS as a theory of diversity. We have only laid the groundwork.

Let us now consider specifics from the ground up. By defining a particle as an organism, with its structural and extra-structural properties, it not only eliminates or even plausibly answers the question 'how did life arise?' But that only goes to show that the question is not as mysterious as it has always been considered. Living creatures are always struggling to avoid death, which is their version of annihilation. It makes sense for this to be their version of annihilation because that is the way we know they cease existing. Particles are also universally reacting with other particles, supporting our definition of an organism that *tends* towards avoiding annihilation. So long as there are particles in the universe, there is the possibility of 'life', whatever that means. Following the Big Bang, the Big Bounce or whatever might have transpired to set our universe in motion, the very appearance of particles marked the appearance of organisms and their struggle to continue to exist. Or as Lawrence

Krauss puts it, stars had to die for us to be born.[6] Organismal Selection shows that it is but a small step between a supernova and life, and as far as particles are concerned it is just life spread out all over the universe again. It might seem weird to call particles alive, so let us just stick with the term organism. I will leave the reader to decide which set of particles are alive and which aren't.

At this point, the theory of OS starts getting predictive. Using probability, we have seen that the probability of annihilation reduces the more the particles merge, mix, conjugate, fuse or combine – whichever term you prefer. The simplest way to acknowledge the validity of this statement is to consider that initially there was one particle, one organism. Another particle then appeared and then there were two particles, which if and after merging, became a new emergent organism, consisting of two primary particles. From two independent particles a compound is formed that shows how the two are now related. This relationship illustrates the formation of a new organism. Another way of looking at it is that from two primary organisms we yield a single secondary organism. If this secondary organism gets annihilated, as nature eventually promises to, the primary organisms will have a near certain chance of still existing, though the secondary one has been annihilated. The two primary particles delay their annihilation by forming another emergent organism by merging. The delayed annihilation is captured by a simple probability function. Viewing particles as organisms that aim to avoid annihilation, OS shows that particles will *tend to seek* these 'mergers'. Organismal Selection therefore dictates that particles will tend to merge with others and, consequently, increase in size. The theory predicts the emergence of multi-particulate organisms from initial particulate organisms. The quarks would merge to form the protons. Hydrogen atoms merged to form helium, and helium merged to form beryllium and carbon. When extrapolated to the current understanding of cellular and multi-cellular organisms, the theory explains the emergence of multicellular organisms from cellular organisms. At this point, it is important to note that OS only predicts the emergence of multicellular organisms but does not dictate how far or how multicellular or multi-particulate the organisms will be. The prediction is more qualitative than quantitative. It is supported by the evolution of multicellular from cellular organisms.[7, 8] It is also supported in particulate physics from the moment the Big Bang happened to the very first seconds of the existence of our shared universe.

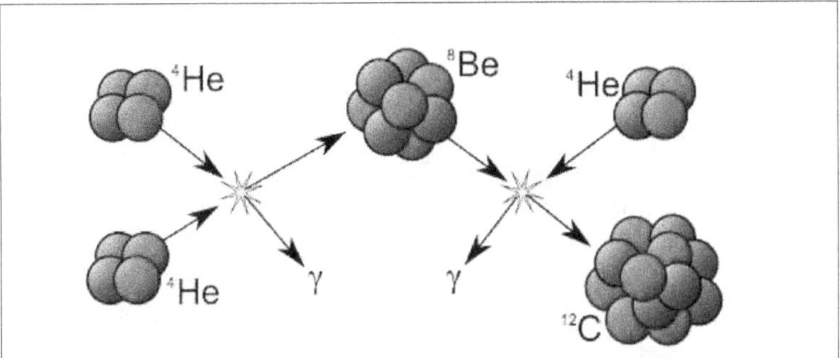

The triple alpha process. In this process, two Helium particles merge to form Beryllium. Beryllium then merges with another Helium particle to form Carbon, the identity element of life as we understand it. Helium particles are also known as alpha particles.

The parallels between an atom and a cell might seem unwarranted. There appears to be an astronomical jump from particles to cells, which practically comprise many particles. At this point, taking the correlation is the most important aspect of the explanation thus far.

The Universe

So far, I have tried to explain one view of how OS is a theory of diversity. This is through the concept of probability of annihilation, which is always gravitating towards certainty, towards 1. Why should it be gravitating towards certain annihilation? It is this question that links evolution with Physics. Physics has a subset of laws that govern scientists every time they want to test their theories. Of all the laws of Physics, there is one that stands unparalleled to all the others. This is the second law of thermodynamics. Broadly speaking, the law dictates that everything tends towards destruction. As I said, the universe has a knack for the boring. Organisms are exciting but the universe just can't take that. This is a gross simplification of the law but it serves the purpose – that there is a natural inclination towards destruction. This destruction can be defined by others as thermodynamic equilibrium or more colloquially, disorder. This state of disorder is known as entropy. Entropy is appreciated by scientists as the measurement of disorder in a system. A more elaborate definition would be that entropy is the measure of how far a system state is from thermodynamic equilibrium.[9] When a system is in a state of thermodynamic equilibrium free energy cannot be extracted from it. Take the example of a cylinder. When it is positioned with the circle at the bottom, it has potential energy, which can be seen when it is tipped to any side. It can fall. However, once it lies on its curved surface there is hardly much energy that can be extracted from it. It can no longer fall. It has achieved equilibrium – a state where free energy cannot be extracted. For biochemists, thermodynamic equilibrium is the most accurate diagnosis of death. Annihilation can be defined by this state, a state of thermodynamic equilibrium. It can be derived from the initial definition we made about the particles – that annihilation is the cessation of existence. When somebody dies their organs are still present. And some still continue to function for some time. There is, however, an irreversible and natural inclination towards thermodynamic equilibrium. It is at this point that the doctor pronounces a patient dead. It is the cessation of existence of somebody as they were known. The organization that constituted a particular person, a rabbit, a hamster or even a dinosaur, has ceased to exist. This is the second law of thermodynamics, which states that for any system, there is a natural inclination towards thermodynamic equilibrium.

However, before anyone dies there is the tendency to avoid death as understood by biochemists or quite simply, annihilation. This takes us to mapping. In the previous chapter, I had highlighted that to appreciate this diversity there is a map that we have to consider. This map includes the universe. The universe as we have understood it has been this vast expanse that physicists and cosmologists study. This is not the universe I intend to highlight. Not in its entirety. Most importantly, I put stress on how the universe relates to the organism. This relationship is standard, constant and robust regardless of how one decides to define organisms.

The map is simple – there is an organism, and its universe. That is the relationship. That is the map. For any particle that exists, it is the organism and anything external to it is the universe. Every organism, therefore, exists in *its own universe*. Let us take this example. In a space-time fabric of a universe that contains two particles, **A** and **B**, **A** is an organism and anything external to it, including **B** is its universe. Likewise, **B** is an organism and everything external to it including **A** is its universe. The map remains unchanged – an organism and its universe.

It quickly becomes apparent that the universe tips towards annihilation but the organism tends to avoid this impending annihilation. On one end, we have an organism that struggles to exist and its universe, which is hell-bent on annihilating it. Its universe is vast and the organism can hardly stand a chance against it. It nevertheless tries. By the simple existence it brings forth its universe, its terminator. But just like the movie, *Terminator*, there is some time before the organism is annihilated. It tends to avoid annihilation as much as it can. This is the relationship that explains more than anything why the organism is independently diverse – why the fundamental unit of diversity is the organism.

What does this mean? It means that this oxygen molecule is different from that one. Physics and Chemistry state that they have the same properties. However, using the map we have just created we see how they are different. In fact, taking the universe as one that exists in the dimensions of space and time, this oxygen is different from that one. This is not the only evidence in support of the map. There is another key assertion, from the same laboratories of Physics, following the ground-breaking work of John Bell.[10]

John Bell was a quantum physicist who was interested in reality or as it is understood in scholarly circles, realism. Realism can be divided into local realism and non-textual realism – but first, realism.[11] Bell argued that realism would be true if the fundamental properties of position, momentum and spin of particles remained unchanged when nobody was observing them. Bell is remembered for developing a set of equations that could be empirically tested to confirm this idea of reality. With the necessary equipment and numerous tests, long after he formulated the experimental idea, it was then shown that this was not the case. These thought-to-be inherent quantities were not constant. It could therefore be inferred that as a first explanation, realism is false. But what of localism? John Bell described localism as the influence of particles that travel at speeds no faster than the speed of light. Experiments set up to validate this definition proved that it was false, which could further mean that the influence of a particle on these quantities could be faster than the speed of light. The other aspect of realism, non-contextual reality, which asserts that the fundamental quantities of position, momentum and spin should be the same regardless of how one measures them, also proved to be false.[12] If these quantities that are used to define one particle from another, which were thought to be fundamental, are not constant, there is no need to attribute a constant probability of annihilation for a particle. Therefore, OS is sufficiently safe if not accurate in insisting the following: there is no such quantity as a fixed inherent probability of annihilation.

While these are the conclusions that physicists then made about reality, that realism is false, OS could offer an alternative explanation. Rather than think that these quantities should be constant regardless of observation or method of measurement, it could be that these quantities are different for each organism. The map that we had fashioned was that of an organism and its universe. My universe is different from your universe. My observations are shaped by how I view objects in my universe. The small differences in how precise I measure the spin of a particle and how you measure it can easily make us consider that the spin is constant. It can easily make us conclude that these quantities are fundamental. For two or more organisms to see the same object, there must be some degree of coordination between these organisms that allows this to happen. But coordination is never perfect. My universe is shifting based on my position, momentum and how I exist in it. Yours too. Coordination can never be perfect. This imperfection could account for the different measurements that were seen after empirically testing Bell's set of equations.

The assertion that realism is false would make a strong argument if organisms shared the same universe. The map we have established says otherwise – an organism has its own universe, unique to itself. Understanding what another organism understands requires high level coordination. For organism **A** to understand that according to organism **B**, the spin of particle **X** is fixed, it must have a perfect coordination with that of organism **B**. This isn't the case. These experiments corroborate our conclusion that the map is still robust – an organism and anything external to it is its universe.

Extrapolating the results of these experiments to our application of probability in OS then shows that we cannot conclude that the probability of particle annihilation is fixed. It is always shifting and eventually gravitating towards certain annihilation. The more the particles fuse, the lesser the probability of annihilation. And the lesser the probability of annihilation, the greater the tendency to avoid annihilation.

Nevertheless, since evolution has long been known to focus on living organisms as we have always understood them, we can further explore and see if OS gets a chance to explain the phenomena that to date is still being passably explained by Natural Selection. We shall soon see that OS and genetic drift are consistent, but OS explains much more, since it is not confined to a genetic understanding of diversity. In addition, OS will show that concepts of systems dynamics, such as non-linearity, self-organization and criticality are consistent with previous theories and even ideas propounded by leading figures such as Stuart Kauffman, Lynn Margulis, James Lovelock, Per Bak, Humberto Maturana and Francisco Varela.

CHAPTER THREE

The Bold Organism

And I love myself (the world is a ghetto with
big guns and picket signs) I love myself (But it
can do what it want whenever it want, I don't
mind) I love myself – **i, Kendrick Lamar.**

It is often stated that one of the best places to capture Natural Selection at work is on an island.[1] Charles Darwin had been an avid collector of insects before his trip to the one island that catapulted him into both fame and scorn. His trip to the Galapagos was no different from Jared Diamond's trip to New Guinea. In the masterpiece that is *Guns, Germs and Steel*, Jared Diamond explained how the setting is perfect not just because it is an island but because of the broad range of diverse species it holds.[2] He then went ahead to show how it shaped the current world situation as we understand it, from the size of the largest cereal to the difficulties in domesticating some of the animals in Africa. Therefore, an island is as good a starting point as it gets to understanding evolution.

Let us picture two organisms of a particular species, in the general sense of the term, present on an island. These organisms can sexually reproduce, and they do, resulting in generations and generations of offspring, long after the original ones have died. It then gets to a point where it becomes evident that the number of organisms in the island has exceeded the carrying capacity of the island – the island's resources have reached a point where the current inhabitants cannot be sustained. The outcome can either be extinction or evolution. Extinction for the weak, those resistant to change or unlucky or evolution for the strong, those malleable to change or lucky. However, this outcome was not apparent to the initial couple in the island.

This is similar to what happens among bacteria when they colonise a surface. *E. coli* are notorious for causing various diseases among humans. Notorious is the word we use when other lifeforms are a threat to us. It is likely that the *E. coli* have a similar word for us, who are hell-bent on annihilating them. So hypothetically, the 'notorious' *E. coli* could invade the anal area in a female patient. Due to the close proximity between the anal and urethral orifice, they can colonise this new surface. They then start reproducing and increasing in number as they find their way up the urethra. The continuous streak of reproduction will eventually hit a plateau phase where the available resources (what the urethra has to offer) do not match the needs of the available bacteria. The *E. coli* would have surpassed their carrying capacity. The result, again, is death (or extinction) or evolution. We are back to initial situation that was not apparent to the first *E. coli* that invaded the urethra.

The all-cutting conclusion that can be drawn from these two examples and many other hypothetical and empirical scenarios is that resources are finite. The finiteness is usually not apparent to the initial bacteria or organism that has a natural inclination to reproduce. The finiteness becomes clear much later. The other deduction we can arrive at when the population is increased is that by having my share of this limited bounty, I reduce the cut for all the others. Initially the bounty seems abundant if not limitless. When the giraffe has a bite of one leaf from the tree, it has reduced the overall number of leaves on the tree and consequently the number of leaves for the other giraffes. While giraffe **A** takes a bite off one leaf, less is left for giraffe **B**. And when **B** takes a bite, less is left for **A**. We can take another approach. While giraffe **A** is *certain* of biting off a leaf from the tree, the total number of leaves reduces and for remaining giraffes, **B** included. Consequently, the degree of *uncertainty* of getting more leaves progressively increases. The reason that taking one leaf increases this uncertainty but does not become apparent to the biting giraffe is partly because the consequence of biting off a leaf becomes negligible if it is spread out among all the remaining giraffes. Say there are ten giraffes feeding off a tree and **A** is one of them. Biting off a leaf means that **A** gets 1 leaf while other nine individually bear the cost of 1/9 of the leaf that **A** just bit off. This is hardly a noticeable feedback to cause the rest of the giraffes to worry. The same applies for the other solitary giraffes when they take a leaf off the tree. Garrett Hardin calls it privatised profits-commonised costs (PP-CC). [3] Giraffe **A** will continue eating the leaves not knowing that it is reducing the available leaves for **B**, and **B** will do the same for **A**. The continuous feeding by the giraffes will eventually exceed the replacement of the leaves by the tree – exceeding the 'carrying capacity' of the tree. The tree will no longer have enough leaves to comfortably feed the giraffes.

The harsh reality of the increased uncertainty will become clear when all the leaves but one, and not more, is left. Let us equip this last leaf with enough energy for the giraffe that eats it to get enough supply to weather the current absence of leaves until the next season. Let us also assume that for the leaf to effect this outcome, it cannot be split in half. How it works is that 'in the spirit of friendship', **A** will not let **B** have it. The same case goes for **B**. They will compete for this final piece. The one that gets it gets to survive the dry season and the unlucky one dies. It then makes sense when Kendrick Lamar, in the lyrics quoted above, says 'I love myself', representing somebody struggling to survive despite the ghetto spirit of nature.

We arrive at something that we obliviously do to ourselves – we annihilate each other without knowing it. By taking this meal I reduce the overall number of meals for the others, since the meal is not infinite. It is finite. By being *certain* of my survival after taking this meal, the overall *uncertainty* increases for the others who did not take that meal. In short, nature annihilates. Nature is red in tooth and claw. It was, therefore, hard for me to marry this concept with Natural Selection. How can nature, which is best known to annihilate, be thought of as 'selecting'? Is that not a self-defeating endeavour? For the organisms that get to have the genetic advantage, did it mean that nature *selected* them? I came to the conclusion that the name Natural Selection did not fit the bill if operationally, it refers to annihilation. These organisms or species that were 'selected' strictly speaking, escaped annihilation. By sticking to this definition, nature annihilates, we can do away with the paradox that the name is likely to bring if one does not understand what nature does. Nature annihilates and those that evolve are those that escape its annihilation. It does not 'select'.

Evolution then introduces a core concept evident in physics – a system and its universe. Nature can be divided into a system and its universe. The invariant map resurfaces. Since the organism is part of nature, it also slowly but surely annihilates itself. It, however, postpones its annihilation by all means necessary. It does this by being open to the universe. The system, which is the organism, is therefore not closed but open. By traversing its universe, it encounters material and energy that it can use to ensure its survival. It tries to maximise its chances of continuing to exist for as long as possible. But to do this, it literally has to be open to possibilities and expose itself to a universe that is hell-bent on annihilating it. As it does this it creates ingenious strategies of harnessing the cruelty of the universe into useful energy for itself. However, these strategies that it contrives do not harness the energy perfectly. Every time the organism converts this energy into useful energy, it continually dumps useless energy into its universe – energy that it will not be able to harness in its future. As a result, it slowly but surely annihilates itself. Nature, that comprises an organism and its universe, annihilates.

For the organism to regularly escape annihilation, it has to be bold in formulating its strategies. The strategies have to be bold because they literally are against the world, or in this case, the organism's universe. The organism can be a whale or an amoeba. It can be multicellular or unicellular. It can be uni-particulate or multi-particulate. Whatever the case, it stands alone. An organism, before eventual annihilation, undergoes an adventurous odyssey, whose achievements it only knows and occasionally celebrates.

What I have just mentioned is in stark contrast with what I have heard from many motivational chants: that once you think it, the universe conspires to bring it into fruition. Your universe does not conspire to make things work for you. Remember, the universe is a ghetto with big guns and picket signs. How *you* escape annihilation, by postponing it, is out of *your* creative strategies. *You* are the creative force that achieves feats that only *you* know since *you* are the only one in your universe. In this sense, evolution is not passive as it has always been accused of being. Lynn Margulis states that nature annihilates but it does not create. I would argue, however, that nature both creates and annihilates. But what do I mean by creating? Remember, nature comprises an organism and its universe. The universe *'creates'* ways of annihilating you, and you *create* ways of escaping this annihilation. It is a *creative* process, tightly coupled. An asymmetrical warfare. Your existence, and that of any organism, is a manifestation of your creativity nnovateon. You are the hero in your own story.

Back when I was in primary school I used to believe that the universe would always conspire to give me what I wanted. This powerful message has helped me, and many others, I am sure, surpass various hurdles. What OS explains is that such a mantra is creativity in itself – a means of surviving in your ghetto universe. It has enough power to either make one sink to the depths of depression, or rethink a suicidal action. But the universe doesn't do this. You do. Your creative and innovative self. This property is evident in all organisms. That a particle continues to exist even after an animal dies is evidence of its bold strategies to postpone its annihilation.

Such chants, that the universe is on your side appear valid in hindsight. The Nobel laureate, Daniel Kahneman, once stated that, a foolish statement turns into an ingenious one in hindsight.[4] I will add that we should always be cautious of what the successful people say. If my theory gets to fit in this bracket as well, be cautious of what I too say. Instead of taking the life out of those powerful statements that keeps very many people going, I'll concede that there is some truth to it. The truth being: *you* are the one that make things work. Also, we have to be cautious with what we mean by truth. What I mean by truth is that the mere existence of an organism, living (cellular) or non-living (particulate), serves as evidence that we are bold in standing against every other thing external to us. I call such strategies bold for a reason – a reason I credit to Karl Popper.

I highly attribute my endearment to science to Karl Raimund Popper. I consider him my favourite teacher even though he taught me posthumously. The first book I ever read by him was *The Open Society & Its Enemies*.[5] I had to confirm that he was indeed a philosopher because the lucidity of his writing and thought process surprised me. That single book challenged my understanding of the society, religion, history and even science. I was hooked. I went on to read virtually all his works, discovering in the process how his ideas shaped science and the philosophy of science in the 20th century. His book, *The Logic of Scientific Discovery*, gave me deep insight into the scientific method.[6] In this book I learnt what Popper calls a bold hypothesis. A hypothesis is bold if it is capable of withstanding the tests of analytical thought and scientific experimentation. Often, its boldness is appreciated by the fact of its hostile reception by the prevailing school of thought. For instance, physicists took a while before they accepted the Quantum theory of reality. Quantum theory, which has prevailed to date, then, is a bold one. This is similar to an organism that is set up against the universe and is still able to devise a strategy to delay its annihilation. This is what I consider a bold strategy.

Bold strategies can account, in a simple way, for how evolution is more than just adaptation. An organism's creative innovation process is often one step, if not several steps, ahead of the universe. It is asymmetric. The mere existence of an organism is evidence of this fact. Its existence means the universe has not succeeded in ceasing (stopping) its existence. Therefore, the organism does not just adapt, but goes further. Specific examples have been captured by the works of Steven Jay Gould and Elisabeth Vrba, who coined the process exaptation.[6] Exaptation is the process where a trait shifts functions different from the original purpose for the shift. The classic example is seen in birds. Birds evolved feathers for the principal purpose of temperature regulation. The feathers can trap air into pockets that insulates them during cold temperatures. Subsequently, they were used for flight, by increasing the surface area and enhancing the streamlining process to suspend the bird in air. The feathers were exapted for flight. Another example is that of the basal fish, which evolved lungs. The lungs might not have been viewed to be beneficial at the time they emerged but they were later moulded from gas bladders, helpful in ensuring buoyancy of the fish. For the one organism that managed to leave the sea, it became moulded to the lungs seen in the terrestrial creatures. The gas bladders were exapted for breathing. If we are to come back home, a subtle, though good example, lies in the story of aspirin.

The Emergent Utility of Aspirin through Exaptation

Aspirin is a pain reliever. Its mechanism of action in the human body classifies the drug as a non-steroidal anti-inflammatory drug. The liver is responsible for the metabolism and mopping up of its toxic products through the biochemical process of glucuronidation. The accumulation of these toxic metabolites results in liver damage and increases the risk of getting hepatitis. How then, were humans able to develop a mechanism to mop out such a toxic chemical compound? The answer lies in the scientific name of the drug – N-acetylsalicylic acid.

Salicylic acid is an important plant hormone that plays a significant role in seed germination, plant growth and development.[7] Man had the initial problem – let's call it P_1 – of finding food. In the event of a futile hunt for animal meat, he would resort to gathering plants and fruits. With the help of several chance events, man was able to solve P_1 by testing several theories and eliminating the plants he found reproachful. Respectively, these were the tentative theory (TT) and error elimination (EE) phases. However, even after finding the useful plants, the concentration levels of salicylic acid in these plants varied depending on the daily intake and food selection by the omnivore. This presented a second problem. Let us call this problem P_2. The problem was, however, not that pressing, since the compound was in small concentrations.

Man thus had ample time to experiment and develop a means of eliminating the toxic compounds, through the TT and EE phases. In addition, various plants that were used in traditional medicine as pain killers contained varying salicylic acid levels. By the time Bayer manufactured the drug the liver had become competent at handling the toxic metabolites.[8] The process of eliminating the acetylsalicylic acid in plants ended up being useful in eliminating the slightly modified compound – N-acetylsalicylic acid. The evolution of a strategy ended up being useful to man when he faced another different albeit related compound. Thus, when man takes an aspirin tablet, the problem (P_2) of eliminating the toxic products of the drug were partly solved by man's previous exposure to salicylic acid. The schema ($P_1 \rightarrow TT \rightarrow EE \rightarrow P_2$) explains, in part, evolution through exaptation.

Exaptation introduces the possibility of acknowledging traits in organisms without knowing the purpose they serve presently or even in the future. Since an organism is present in its own universe, this is one of the possible explanations for emergence of unique traits. At times, the organism might even die before one is able to appreciate the role of some of its peculiar traits. I for one, have what I consider a rather rare trait – the main transverse creases on my fingers are not typically three on all fingers. My left hand has five creases on my thumb, four creases in my middle three fingers and three creases in my small finger. In my right hand, my middle finger has four creases as well as my thumb, and the other fingers have three creases. I have never understood why I have these many transverse creases. If only I could donate one or cash it. This interpretation is a step away from explaining arrival of the fittest, which I believe OS is well equipped to do. In this sense, these traits are also bold strategies, which manifest following DNA translation. We can therefore safely conclude that these unique structures can be considered to have stemmed from bold instructions.

Why Bold Instruction?

Deoxyribonucleic acid (DNA) has been likened to a blue print, even though a blue-print hardly allows room for changes such as DNA mutations. A better analogy is a recipe, which can be tweaked here and there.

The recipe contains instructions on how a meal should be prepared. Numerous studies have tentatively deduced that DNA contains instructions on how the body should be built. When the body is under attack by the universe, the DNA contains some instructions on how to respond so as to preserve the integrity of the body. Therefore, a bold instruction is an apt way of describing the strategies that organisms implement to escape annihilation at the genetic level.

Before some of these strategies are wielded, DNA has to churn out a series of instructions for the cell, organ or body. For instance, when a resident by the seashore decides to go hiking up a tall mountain, the high-altitude conditions prove daunting, different from what the body is accustomed to. This harsh environment stimulates the kidney cells to manufacture a protein known as erythropoietin. This protein is only manufactured by the cell after these instructions are churned from the DNA. The erythropoietin then escapes through the blood into the bone marrow, to stimulate cells that are yet to specialize, into becoming red blood cells. In this sense, a set of instructions have been executed. A bold instruction then gives the seashore resident a chance of surviving the harsh relief climate atop the mountain.

Conventionally, it is believed that DNA archives these sets of instructions. Such a conclusion excludes the stimulators that initiated this instruction, begging the question – did the instructions come from the DNA or from the stream of molecules preceding the transcription of the DNA? The consensus lies in the idea that reproduction facilitates the transmission of traits through the duplication of the DNA, giving the scientist reason to believe that the instructions are archived in the DNA. This is a conclusion I am not willing to *fully* accept, even though it bears a lot of weight. A model as simple as that of genes stimulating change does not account for the complexity of organisms.[9, 10] What I would yield to is a concept known as self-organization. Self-organization incorporates all the actions preceding, during and proceeding from DNA transcription. It does not single out instruction in one particular archive but in the entire organism. Self-organization equips an organism with the ability to evolve in the sense of avoiding annihilation. It does not have to stem from the DNA's instructions. As self-organization exists in the presence and more importantly, the absence of the idea of DNA or RNA shifting, drifting, turn-over or mutation, I wager it is bolder than the idea of DNA transmission of instructions. Self-organization is another concept that I will use frequently in the course of the book, as it is instrumental in appreciating particles and unicellular and multicellular creatures as organisms. It will also be pivotal in explaining the arrival of the fittest, later on in the book. But before that, here are some examples of bold strategies executed by organisms that are worth mentioning. Let us begin with humans.

We have already seen how high altitude stimulates the production of more red blood cells. This, however, is not the most threatening problem facing human beings. Arguably, one of the biggest threats to humans stems from vectors, the most lethal being the mosquito. According to World Health Organization (WHO) mosquitoes result in the death of millions of people every year.[11] This is an estimate of registered deaths every year, often from malaria. The deaths are several integers greater than the estimated number of deaths from the grievous Hiroshima and Nagasaki incident. Nevertheless, we have been unable to come up with permanent interventions against it. Strategies range from clearing bushes and draining puddles to using insect-treated mosquito nets and insecticides. An unappreciated facet in evolution is that of using *any* material to ensure organismal survival, such as an insect-treated mosquito nets. It does not have to come from the DNA. The rule is to use whatever you can to avoid annihilation.

If we are to pause for a moment to examine the interaction between man and the mosquito, it becomes clear that the mosquito, like other organisms, employs various means to escaping annihilation. As a vector of the disease-causing agent (*plasmodium species*), the mosquito needs to find ways of surviving. In particular, it is the female anopheles mosquito that transmits this agent during a blood meal. The female anopheles mosquito needs to take a blood meal for it to nourish its developing offspring. But this involves the risk of being annihilated by man. On the other hand, it favours man to subject itself to pain temporarily by slapping our arms or other parts of the body with the intention of swatting mosquitoes, than contracting the disease that the vector brings. It seems better for humans to swat the mosquito now than to put up with the irritating buzz. This also reduces the chances of contracting other diseases transmitted by mosquitoes other than malaria. Therefore, after several generations and death episodes, the remaining group of mosquitoes should be more likely to strike either when the person is unaware, or when the accuracy of this annihilation machine (you and me) is reduced. This could be the reason some mosquitoes mostly strike when there is reduced human vision, that is, when it is dark, or when humans are asleep. That is not all. Mosquitoes need to access specific channels of food from its annihilator. Pick the wrong channel and you die, pick the correct one, and you survive. Mosquitoes must, therefore, be equipped with a mechanism to detect arterial from venous blood supplies. The ability to distinguish between the two vessels distinguishes those that escape annihilation from those that do not.

For the strategic positioning and close monitoring, it might be apt for the mosquito to rest within the environs of man, either inside the house (endophillic) or outside it (exophillic). Even more interesting is the strategy that I have always believed to be beneficial – the flying buzz it makes. By having such a distinctive noise, it is able to detect if the meal is awake or not and thus better situated for striking. These, among other employed strategies, can be considered bold in that they have displayed exceptional creativity in escaping doom either intentionally or unintentionally. The noise the female mosquito makes is an example of how organisms more than adapt to their universe.

Much more creative, however, is the agent itself that causes the malarial disease. By the mere attachment to mosquitoes, *Plasmodium species* hitch hike on all the tricks crafted by the female anopheles and don't stop there. They ensure that they always shift their molecular configuration, so that the anti-protozoal medication that man uses with every generation, gets weaker. It is one of the reasons why getting a malaria vaccine has eluded mankind for the longest time. The same can be said about the influenza virus, and why there is a regular rollout of influenza vaccines.

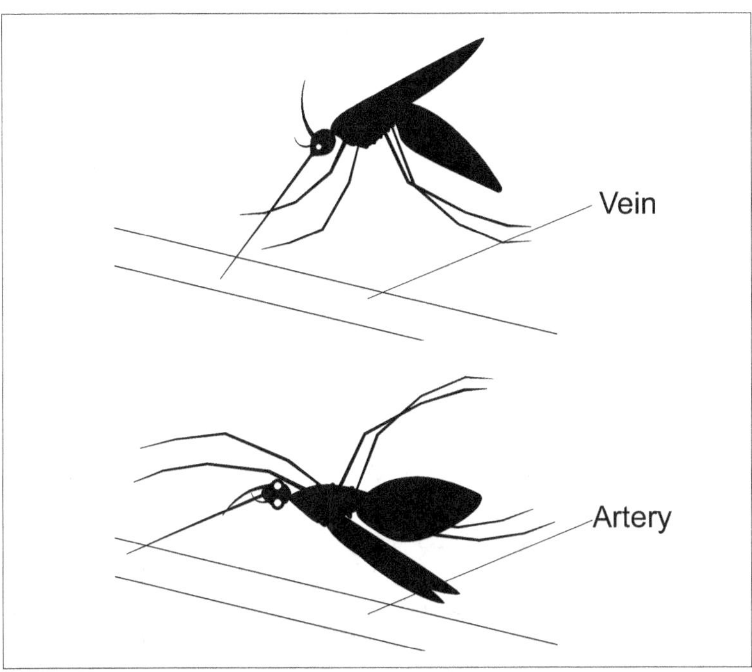

Mosquitos during a blood-meal. High pressure in arteries can be deadly to mosquitoes compared to the low blood pressure in veins. Mosquitoes have evolved a sensitive proboscis to distinguish a vein from an artery.

This property of molecular reconstitution still falls under the systemic term – self-organization. The word has stuck its head out once more. Self-organization is a dynamically-enhancing property that arises in a system without prior planning. For instance, the mosquitoes that attack human beings at night are likely to end up being more prevalent than those that strike during the day, when humans are awake. Giraffe **A** slowly but surely destroys giraffe **B**, giving the tree enough time to grow its leaves once the giraffes are dead. Self-organization plays an indubitable role in the patterned behaviour in all organisms.

The extents of self-organization can be appreciated by the strategies that humans have yielded to when addressing their problems. For instance, eyesight. The normal human vision is bound to vary, either from childhood or in late maturation into adulthood. Purchasing spectacles can correct this defect much faster and more easily than having to wait for reconstitution of the body's molecular properties. All the while, the organism is racing against a universe that is out to eliminate it. Spectacles equip this person racing against time with an acuity that another acquired through a robust genetic makeup. With reference to vision, these two people (acquired acuity and inherent acuity) would be ranked equally. There would then be a pattern of people purchasing spectacles to achieve visual acuity, taking the easier, cheaper and faster option. In short, OS shows that bold strategies do not have to come from the DNA. This is well reiterated in Talib Kweli's song *Get By* – an organism will do anything just to get by.

Popperian Schema

All these strategies are appreciated only, if not largely, in hindsight. They must have somehow emerged before we even got the chance to acknowledge their bold status. For them to emerge we will focus on two aspects – the problem and the bold solution. First, the problem. Up to this point, it has become clear that every organism is literally against the world, and in cosmic levels, against the universe. This is even captured by the fact that they have an attributable probability of annihilation. Consequently, there is also a tendency to avoid annihilation. The probability hints at the presence of a problem – the need to avoid annihilation. The mere existence of an organism shows not only that there is a problem, but that there is a solution that it has formed – although, as we'll see, the solutions are less than perfect. This is so because annihilation is certain later in the life of the organism. According to the modern synthesis, solutions are largely genetic. Genetic solutions, however, can be pointless to a kitten if a tree lands on it. Since we are already talking about kittens, let's take on a feline perspective.

Felines are territorial creatures and they have various ways of demarcating what they consider 'theirs'. Among male lions an alpha marks its territory and knows the scent of the lions in its pride. If another male lion crosses the territory, it risks getting into a brawl with the owner of the territory. Let us, however, imagine that the two males get into a brawl. If the resident alpha dominates the trespasser the trespasser might opt to run and thus escape death. It might then find another territory and dominate the alpha male. The trespasser, therefore, escapes its doom while incidentally securing its status in another territory.

The trespasser might have won the fight in the second territory, but from game theoretical studies by Maynard Smith, this is not the evolutionarily stable strategy.[12] Getting into a marked territory exposes the lion to a potentially fatal injury and is, thus, a waste of energy. Initially, energy is wasted in seeking a territory. It then sub-optimally faces the reigning alpha in that territory, having already wasted a portion of its energy roaming in search of a territory to conquer. A simpler strategy is marking your own territory and only allowing trespassing lionesses to become a part of it. It is the more stable strategy that gives it a better chance of avoiding death. This evolutionarily stable strategy is behavioural.

A dissection of the success of any such strategy could reveal that it might be genetic. In the case where the trespasser ran away from the first alpha it encountered, it could have been that the alpha was larger. The genetic endowment could not have favoured the trespasser with size but it could have equipped it with speed. Alternatively, running away could be a behavioural strategy that is not directly linked to genetic endowment. Strategies can therefore be structural or not, genetic or behavioural. They final outcome can, therefore, be attributed to genetics, behaviour, both or it could even be circumstantial.

A reductionist argument might link the ability to escape doom to some subset group of genes, but even in a cub litter of similar genes one might get burnt by a forest fire, just because it was ill-positioned when the pride started running away to safety. The genes, or even the genotype, in this case, could not help. The lion cub was just unlucky. The strategy by the parent, however, can be considered bold because it ensured that it had more cubs, so that at least some can escape annihilation. The unlucky cub could have decided to jump onto the parent and cover more ground because of its mother's bigger strides. This could be interpreted as behavioural, even though the cub could be genetically weaker than the rest in the litter. On a grand scale strategies are all similar: they intend to get the organism into safety and far from doom.

Hence, with the universe pegged against you, you will always have problems. Your luck and how you handle these problems is what counts. Moreover, these solutions do not appear in groups. They appear individually, then are later adopted by others, casting the smoke-screen appearance that they are from groups. Probability shows us that there is a greater chance of a single person being called 'Boboshanti' than two people, and the probability is higher for two people to bear that name, than three. Similarly, the probability of a bold strategy emerging from an organism is much higher than from several organisms in a group. The emergence of the fittest has a more probable chance in an individual organism than in several organisms. This tight coupling between an organism creating emergent strategies and its universe manufacturing new problems can be captured in what I call the Popperian schema.

The Popperian schema, as you might have guessed it, was borrowed from the works of Karl Popper. Karl Popper was so interested in evolution that he discovered one flaw in Darwinism – the passivity that earlier I hinted about. He noted that this was a mistake transmitted by the prevailing paradigm, that is, the bucket theory of knowledge.[13] According to this theory, every organism is likened to a bucket, which needs to be filled with knowledge. How one fills their bucket with knowledge was simply through observation. Popper, however, found this theory absurd. To prove this point to his students he would always ask them to observe, to which the students would ask: observe what? This was all he needed to explain to the students his view, that knowledge is created by the organism through an active process. You have to have a 'net' for you to know what to observe. The organism must first create this 'net', which is a hypothesis, then test it out in its universe.

The 'net' is critical for organismal survival. Every organism is always bombarded by a multitude of stimuli that it simply cannot handle all of them at once. It has to select. It has to establish a filter, a net, for selecting. This is an active process of acquiring knowledge. Since the hypotheses it creates are its own, the information it gathers is known only to itself. This is the means through which generation of knowledge starts. The knowledge that an organism acquires is solely its own, since everything external to it is its universe. In a nutshell, it writes its own story.

This is in contrast to the experiments by the Nobel laureate, Ivan Pavlov. Pavlov is said to have conditioned dogs to salivate when hearing him ring his bell.[14] This interpretation highlights a passive role of the dog, ignoring the active role of the dog had in the experiment. The more accurate interpretation is that of a dog minding its own business, creating a hypothesis that trapped that singular stimulus in the environment – a bell ringing. The dog could then have made a hypothesis, say, 'the bell might lead me to another dog that I might befriend.' It then decided to seek out what the bell might be indicating, if it supported its hypothesis. If it found another dog, who happened to be friendly, the dog would have learnt that its hypothesis was valid (as it predicted). If the ringing of the bell did not lead to such a dog, it would have rejected the hypothesis or modified it. In the case of Pavlov's experiment, the dog is likely to have modified the hypothesis from – 'the bell leads to another friend' – to, 'the bell leads to food'. The next time Pavlov rang the bell, it might have wanted to test the hypothesis again – 'will the bell still do the same today?' The bell validated the hypothesis that the dog created, not what Pavlov did!

The same Popperian principle could apply to Pavlov – he formulated a hypothesis that went thus: 'if I ring a bell, maybe the dog will get interested.' If the dog was simultaneously noticing pheromones of a nearby dog on heat, it would have ignored Pavlov. Pavlov would have had to reject or modify his hypothesis. This understanding gave Popper enough courage to coin the idea of active Darwinism. He thereby showed how evolution is not a passive process but an active one. The organism actively learns by solving problems whenever its hypotheses are supported in its universe. Nonetheless, we have already explained how the solutions are less than perfect. In the same way that one feature might have unforeseen benefits, one solution often births another unforeseen problem. The organism's active problem-solving further confirms how in general, nature annihilates by creating unforeseen problems.

According to OS, one of the ways a particulate organism seeks to solve its problem is by merging with another particle. It can be assumed that the single present problem a particle can face is the continuous shifting of its probability of annihilation towards **1**, towards certainty. Henceforth, a particle can seek another to bond and form a stable relationship. This strategy ensures that the probability of annihilation of the emergent organism, as we have seen, reduces. The technical term for such a 'merger' is *complementarity*.[15] For particles to merge they have to be complementary. Complementarity can be steric or electrostatic. Steric complementarity is seen when one particle fits into another seamlessly, like Lego blocks or pieces of jigsaw puzzle, since the shape of the edges complement each other. It is at times equated to the lock and key idea, or hand to a glove. Electrostatic complementarity has to do with electrostatic forces. The positively charged particle attracts the negatively charged one. Complementarity, as a concept, accepts these two mechanisms. Organismal Selection, however, includes much more than that. Later in the book we shall discuss how other forces, such as gravity, play a role in the merger. For now, we can take the idea of complementarity as a merger between two particulate organisms.

Bearing in mind that complementarity requires a seamless and a worthwhile effort in merging with another particle, there is another problem that the particle has – finding a *complementary* particle. This is not usually instantaneous. Since everything external to the organism (particle) is inclined towards destroying it, the organism is more likely to end in failed than successful mergers. Even sadder is that once it has found one it does not solve its problem completely. It only does so partially because while it has participated in the formation of a new entity, it has also created another problem – a new probability of annihilation. Its existence is now significantly a function of the emergent entity, the emergent organism, and how well it plays its cards to avoid its doom. A part of the burden is shifted from the original particle to the emergent organism, the new compound. This new organism has to find ways to avoid its annihilation, saving the original ones much trouble. However, it takes away from the original particle a portion of control over its existence, presenting a problem that it cannot address. The latter has to trust that the emergent organism will be creative enough to prolong both of their lives or existences. In systems lingo, this is somewhat similar to a classical archetype known as: shifting the burden to the intervenor. The original particle shifts the burden of surviving to the emergent organism after the merger. It fosters dependence. Thus, while the merger partly solved its initial problem – a constant shift towards certain annihilation – there is also an emergent problem created by the merger – reduced control and dependence. To this point, we can form a schema.

A particle moves from the first problem, appreciating that it has two fates: inbound annihilation and the small chance to avoid it. It is inclined to avoid annihilation by creating a solution such as seeking another conjugate particle, one that it can form a new entity with or be complementary with. Before it gets one it is met with a lot of failures, to be avoided, or errors, to be eliminated, in order to find a more *cost-effective* relationship. It is a cost because it has to sacrifice some degree of control over its actions. It is also effective because it delays annihilation. When the particle finds one suitable partner the probability of annihilation reduces and its overall tendency to avoid annihilation increases, but consequentially creates another problem. It loses substantial freedom and is forced into dependence. Additionally, an increase in its tendency to avoid annihilation shows that the problem is not completely solved, hence there is still a problem even after the solution of a merger is offered. Simply put, the schema is:

$$P_1 \rightarrow TT \rightarrow EE \rightarrow P_2$$

Where P_1 is the first problem, **TT** the tentative theory (solution to the problem), **EE**, the error elimination phase and P_2 the unforeseen secondary problems that form after solving the first problem. I refer you to the active process of knowledge formation that Popper elaborates, which correlates to the tentative theory phase in this schema.

This is the Popperian schema. With this model Popper was able to demonstrate how knowledge begins and how it grows. While his initial focus was largely on theories and how they could be deemed scientific or not, this schema proved robust when extended to other disciplines.

As such, the impact of this schema is significant to evolution by considering the interaction between an entity that struggles to avoid doom and a universe that is hell-bent on destroying it. The bold solutions, which in this case are the tentative theories, capture the creativity of the organism. The *tentative* theory is aptly described because the solution is only temporary. In a universe that annihilates, all solutions have expiry dates. By formulating hypotheses and theories throughout their entire lifetimes, organisms are therefore knowledge generators. The lifetime of an organism is an entire lab operation, a hostile lab operation, an experimental set up.

A more accurate version of the schema would be:

$$P_n \rightarrow TT \rightarrow EE \rightarrow P_{n+1}$$

Where **n** is the nth number of the problem. A temporal version of the schema, which captures the periodic formation of the problems would be:

$$P_t \rightarrow TT \rightarrow EE \rightarrow P_{t+1}$$

Where **t** is time.

It is important to note that this relationship is not linear, but nonlinear. The probability change shows this. For instance, the merging of two particles with an equal annihilation probability of 0.5 becomes 0.25 through multiplication. Hence the non-linearity. By forming a cost-friendly merger with another particulate organism, the new emergent organism has a smaller probability of annihilation. It translates to a delayed annihilation for the constituent organisms. The complete annihilation of the emergent organism *and* the constituent organisms is then captured by multiplication of the probabilities. From this simple schema we then see that cooperation appears as the better and evolutionarily stable strategy compared to competition.

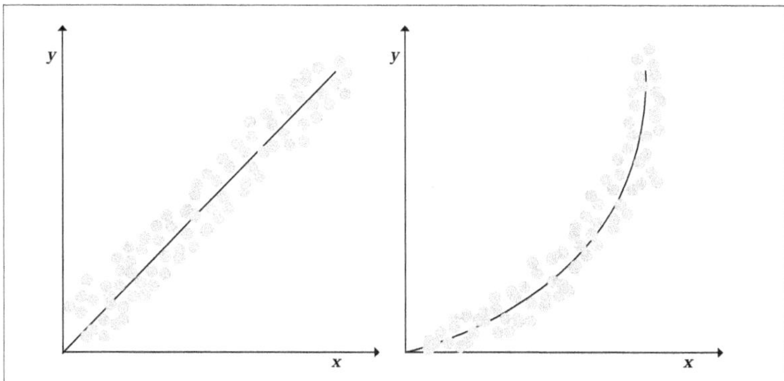

Linear versus Non-linear graphs. Non-linear graphs take a faster progression along the y-axis relative to the x-axis.

It is difficult to imagine that a particle can bear the same properties as those which we attribute to living organisms. As a medical student, I always grappled with the concept, finding it easier to think of 'life' according to biological dictates. Particles, nevertheless, fit in the Popperian schema, and this paradigm shift has since put me in the state of wonder I now hope to bring to you, dear reader.

The map that we have pictured is still robust. We have the particle and its universe. In its draconian universe, the particulate organism is always burdened. In the active process of solving its problems, the schema shows that it generates solutions and in the process, knowledge, as the organism understands it. The solutions, like theories as we have grown to understand them, only work for a certain time. There is a continuous dance between the organism's continued existence and a universe that continuously attempts to annihilate. Functionally, the organism has a property of *autocatalysis* – 'auto' to mean self and 'catalysis' to mean continued existence – and for the universe, that of *inhibition*. The organism is never bigger than its universe, so it occupies a small space. Its universe is much larger and occupies the rest of the space in nature. The organism is localised, while the universe is widespread. Combining these properties they contain, we have an autocatalytic phenomenon known as the organism in a tiny location, and an inhibitory phenomenon, known as the organism's universe spread widely and external to the organism. Briefly put, nature would have a localised autocatalytic phenomenon and widespread inhibitory phenomena. We have independently arrived at a process that the English mathematician, computer scientist, logician, cryptanalyst, philosopher and theoretical biologist Alan Turing discovered. This process is what John Gribbin called the Turing Phenomenon, in his insightful book, *Deep Simplicity*.[16]

From the hefty titles preceding his name, it is clear that Turing was interested in many things. He is famously remembered for leading the team that cracked the Enigma machine during the World War II. However, Turing's curiosity was not just limited to machines. His interests stretched into Biology where he also wanted to crack the *enigmatic* mechanism behind cellular morphogenesis. Cellular morphogenesis is the process that a cell undergoes before it becomes specialized, hence taking certain spatial and functional features. These features are specific for different cells. Turing published a paper on this in which he discussed a process that he called *Reaction-Diffusion (RD)*.[17] The RD system comprises two interacting entities: a localised self-propagative phenomenon and a wide-spread inhibitory phenomenon, which we shall call the Turing mechanism. This mechanism creates various regions surrounding a developing cell known as morphogenetic fields. A morphogenetic field moulds a particular naïve cell into a definitive one, which is a crucial step before organ formation. (By morphogenetic field, I do not refer to the one popularised by Rupert Sheldrake. I refer to the type that can be tested in laboratories and modelled by computers. This later type was advanced by Brian Goodwin, who was interested in improving Turing's ideas.)

Morphogenetic fields contain properties of self-organization, a term we have previously encountered, and which can be seen in a developing embryo. When cells are transferred from their initial morphogenetic field to a different place of the body, and hence another morphogenetic field, it has the potential to form new definitive cells and organ, different from the initial morphogenetic field. In cellular organisms, self-organization capabilities define outcomes that are largely independent of the genetic lineage and, as such, display features lacking a central form of control. For example, if cells of the developing liver were transferred to the limb bud, they would turn out just like typical cells of limbs. Self-organization ensures a fairly consistent outcome. The plausible conclusion we can deduce from such a mechanism is that genetic transcription and epigenetic processes could just be self-organization manifest. Rather than think that the 'genetic blueprint' formed the organism or an organ (which is part of the organism), it could just be that self-organization properties yielded a fairly consistent outcome.

Since its inception, self-organization through reactive diffusion (RD) has been ignored until recently, when RD models have been noted to predict and yield incredibly consistent results. These and several other optimised models have been incorporated in the new field known as *Mathematical Biology*, which create a series of equations that have immense predictive and explanatory value.[18] For instance, RD models have been deemed useful in explaining some features in animals such as the orange-white patterns in the clown fish, the stripes of the zebra and even the leopard's spots. Using computer models, Goodwin has even shown how *Acetabularia acetabulum* gains its unique form.[19]

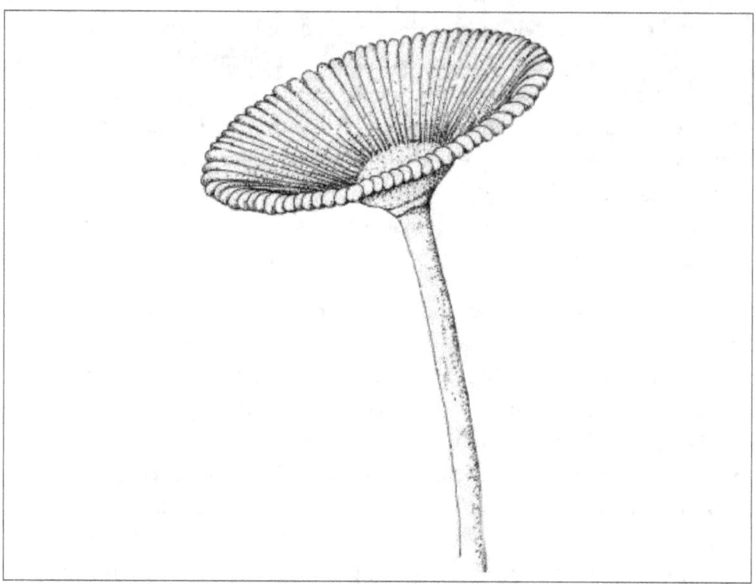

Acetabulum acetabularia. This large cap-like structure is a single-celled organism displaying the complexity seen in multi-cellular organisms. As proponents of self-organization and in particular, structuralism, Goodwin and his team were able to develop models that predicted the shape of *Acetabulum acetabularia*.

Criticisms have, however, been hurled at the *inefficiency* of this system. The first is that although the model is mathematically fluid, it lacks a biological support. What this means is that the model is not supported by evidence gathered on how genes work. It becomes difficult to imagine an elegant equation capturing the erratic picture of gene networks and ecosystems. Furthermore, following from Natural Selection, genes ought to have a causal basis, which is: exploit the organism until it reproduces. A mechanism based on interactions of particles distances the 'mathematically fluid' model from such a causal basis. However, keen analysis shows that this argument does not tackle the question of how 'efficient' the model is or lack thereof, but rather how the model does not completely agree with that of Natural Selection. A mismatch does not imply inefficiency.

The second argument has to do with the constituents of this RD model otherwise known as morphogens. Since Turing was interested in cellular *morphogenesis*, he named the reagents in his experimental system *morphogens*. One morphogen would be self-propagative, that is, autocatalytic, and the other inhibitory. His insight was that the interactions of the morphogens ought to eventually result in the complex process of cellular morphogenesis. The RD system should then possess one of the simplest, most stable mechanisms that should be evolutionarily favourable, as evolution is notorious for selecting the easiest, most stable of mechanisms. Opponents insist that while such an outcome is agreeable in theory, it is not observed in reality, where more than two baseline genes are always at play – these are maternal, gap, pair-rule and segment-polarity genes. By correlating genes with morphogens, the conclusion was that the Turing mechanism was not in concert with reality.

Following OS, I tend to think otherwise. The map of organism and its universe is one that incorporates the kind of models used in the RD system. Organismal Selection captures the organism as the localised phenomenon, and its universe as the widespread inhibitory phenomenon. Our definition of a particle as an organism and its interaction with its universe further incorporates the various chemical reactions that characterise the RD system. Since biology does not consider particles as organisms, it makes sense for some biologists to think of RD models as the kind that lack biological bases. The interactions of elementary particles do not hint at the eventual goal of reproduction. As far as OS is concerned, from the Turing system springs out a causal basis different from the one fashioned by Natural Selection. While Natural Selection asserts that organisms are pressured to reproduce, Organismal Selection states that they are pressured to avoid annihilation. When particles interact, they try to seek solutions to postpone their annihilation. Furthermore, mathematics has been shown to play a role in physics for a long time. However, it was not excluded as a tool of interpreting physics concepts simply because it lacked a basis in physics. The RD system cannot be disregarded because it lacks a biological basis. It is an alternative interpretation, one capable of enriching our understanding of reality.

As for the second argument, it has prevailed due to the body of evidence from genomic studies that eclipsed the morphogenetic field hypothesis. Scientists are ideally supposed to carry out experiments and follow where the results lead them even though it might be at odds with what they believe to be true. The fact that the RD model is able to account for various traits that other models can't is not something that ought to be ignored. It should rather be explored if not modified to yield better results. One way of doing just that is considering any constellation of particles as a variable number of interacting organisms. One therefore does not have to discard the model just because it does not fit the four basic genes seen in living organisms. In fact, if each of the four genes were considered to be organisms, then our map should dictate how each relates with its universe. For instance, if we consider a maternal gene as an organism, its universe includes the segment-polarity gene, gap gene, pair-rule gene and any other particle that is not *that* maternal gene.

Fortunately, a new wave of scientists has taken up the latter responsibility and are creating what they deem to be better models. These models have not just stemmed from biology, but from other fields. Lewis Wolpert, a civil engineer in training, has shown how a similarly 'fluid' explanation, known as positional information, plays a role in biological systems.[20] Wolpert intuited that genes do not know where they are in a cell unless they interact with their immediate neighbours. This is the idea behind positional information, a mechanism that points towards the influence of the immediate surrounding of an organism, almost similar to but different from Turing's idea of autocatalytic phenomena. We have a mind to thank James Murray, Brian Goodwin, Lewis Wolpert and other like-minded scientists for utilising mathematical concepts to explain biological mysteries. Organismal Selection generally, if not strictly, supports their assertions. Considering the Turing mechanism as it is, localised autocatalytic phenomena and widespread inhibitory phenomena has far-reaching consequences, an insight that we can begin to appreciate with a deeper scrutiny of the particle. So let us go back to the particle.

Organismal Selection elaborates how a particle is equipped with key qualities for it to be considered an organism, and hence worthy of being considered in evolution. These qualities can be likened to some aspect in the history of quantum mechanics. For a long time, the view that light was a wave had dominated the scene until two giants stood to question this concept. Albert Einstein and Max Planck proved that light can be considered not only as a wave, but also as a particle.[21] While the world was still grappling with this finding, another giant stepped into the scene and was able to convince the world that particles also had a wave attribute to them. This person follows from the lineage of Louis XIV as Louis de Broglie.[22] Thus, for the longest time, light, which was thought to be wave-like, obtained a wave-particle duality. Matter, which was thought to be particulate, also had a wave function attributed to it – matter also had the wave-particle duality. In the same spirit of dichotomy, OS has also given particles a duality that incorporates particulate matter into the field of evolution – structural and extra-structural attributes. The structural comprises the physical matter, the particle, the cell, the multicellular being. The extra-structural is the annihilation probability and consequently, tendency to avoid annihilation. In quantum mechanics, waves are expressed as probabilities. In OS, the particle has this probability of annihilation. From this attribute, a particle can be personified, as it tries to escape death. As we have seen, light was thought to be a wave, until a scientist questioned it, claiming that it could also behave like a particle. The electron was also thought to be a particle until another scientist thought that it could also behave like a wave. Similarly, I would want to ask: why should we not consider a particle as an organism? From the probability of annihilation and consequently, its tendency to avoid annihilation, a particle can be considered to bear cognitive properties. For it to escape annihilation, it should at least be aware of its problems and the need to escape them. And indeed, it is likely that this is the case. How? Let us take a brief detour to elaborate this possibility.

It is generally accepted that particles always if not often takes the simplest route. This might be equated with the path of least action or the path of little resistance. It is a nice solution that we can take, but then again, nature does not care whether we notice its simplicity or its complexity. In fact, we do not know whether it cares that we know or don't know. Why we decide to focus on simplicity is only a confession of our inferiority in understanding nature, by generating simplistic theories with significant explanatory power. Take for instance, the bold solutions an organism formulates in its own universe that get validated by another organism, and its own universe. Einstein, in his universe, acclaims that Planck's theory of the particulate theory of light is valid. Surely, creating such theorems that are fairly invariant when confirmed by another organism is a testament to the bold nature of these scientific theorems. It, therefore, makes sense why Feynman considered discovering these laws himself, independently, to test their validity. The high level of precision that is independently tested and verified by every organism is a testament to the validity of the hypothesis. This is one way of looking at it. The other is that our behaviours are so coordinated, that there is little chance of our observations differing. This I believe is equally valid, and I shall return to it later on in the book. For now, let us get back to the light particle.

It has become clear to scientists that considering light as a wave explains, to a good approximation, how it behaves as particulate matter. Classical physics states that light moves in a straight line. Quantum mechanics quite literally bends this assertion a little, stating that light takes the path with least resistance, the one *closest* to a straight light. To understand this, there is a concept about waves known as interference. When wave crests (the top parts of a wave) are close to each other, they interfere with each other, in a supportive way, resulting in a wave that is much bigger than individual waves. When two wave crests positively interact this way, the process is called constructive interference. When a wave crest and a wave trough (the dips in a wave) are close to each other, they interfere negatively and destroy each other. This process is known as destructive interference. How then does the light particle know up-front which path to choose? According the Richard Feynman's theory, the sum-over-paths theory, it tries them all. After this trial run, it considers the paths with waves that constructively interfere. The path with the greatest constructive interference is the most favourable and the ones that the light particle and wave follows. The most favourable path is the one closest to a straight line. Following these quantum mechanical experiments, it turns out that the straight line is a good approximation of the constructive interference paths. However, it is easier to appreciate the wave pattern of a small particle than that of a larger one. This is summed up by the work of Louis de Broglie, where a large object, such as you and me, express their wave-like property less than a small particle, such as light.

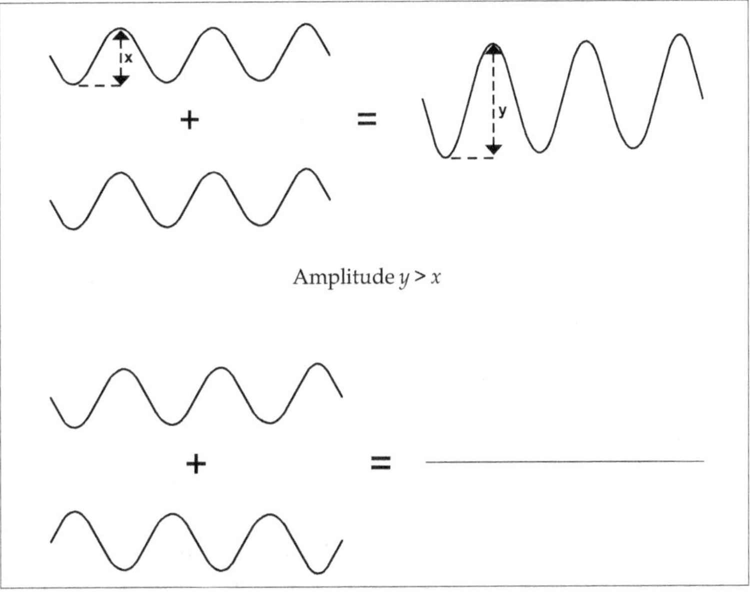

Constructive and destructive wave interference. In constructive interference, when two waves merge, the resultant amplitude is greater than the simple additive process of the two. In destructive interference, the two amplitudes obliterate each other.

How does it tie with the cognitive processes that OS ascribes to particles? The problem that the light particle is faced with is choosing a path that suits it, which would not annihilate the particle; the one where particles do not destructively interfere. It, therefore, explores its options – by trying all the paths, in the manner implied in Feynman's sum-over-paths theory. It only opts for the path that constructively interferes. Only then does it take this path of least resistance, as the other paths are destructive. Simply put, light particle was avoiding destruction or in other words, annihilation. Light does appear to have cognitive capabilities. It seems to be aware of the problem (destructive interference) and solves its problem by literally running away from it, seeking constructive paths. It is, therefore, a matter of convenience that we attribute nature to using the simplest and shortest routes, because in actual sense, it tried all the routes and found them bent towards annihilation. It makes sense that for the particle to avoid annihilation, it has to avoid the destructive paths.

In fact, in a bid to entertain ourselves, it can be thought that the wave-like trait of particles is the problem-solving bits that pave the way for the particulate bits. This is speculative, but then again, all ideas, great or small, start with speculation. The OS theory therefore explains why organisms, as we understand them (biologically and in their particulate nature), opt for the strategies that take the least effort, as seen in the example of light particles (and their sidekicks, the waves). If at this point you still are not caving to the possibility of particles having capabilities such as solving problems, much like all living organisms, I understand. I would be lying if I were to say that I have completely shaken off that belief. The part of me that can't shake off its twenty-year history with biology still believes that considering particles to be organisms is an absurdity. The far-reaching consequences of such an interpretation nonetheless make me consider it as a possibility, a possibility that I would like to share with everyone. It is the possibility of showing how the various disparate academic fields separate similar ideas. The elusive development of a framework that explains arrival of the fittest, one of the weaknesses of modern synthesis, is a classic example. By borrowing leaves from various disciplines, this is a concept I believe that OS captures rather well, as we shall see.

Statistics and Symbiosis

I do what I do and you do what you can do
*about it – **A Milli, Lil Wayne***

When developing this theory, the one aspect that bothered me was the precision. In subjects such as physics, precision is highly regarded and proves to be a testament of the validity of the theory. Newton is famous for his equations of motion and gravity; Maxwell for his field equations; Einstein for his equation of gravity and controversial cosmological constant; and the pioneers of quantum physics, like Erwin Schrodinger, Werner Heisenberg and even Richard Feynman, all generated precise concepts. In evolution, proponents of various theories also inclined toward mathematical precision. Natural Selection obtained a statistical interpretation thanks to the works of Ronald Fisher, J.B.S. Haldane and Sewall Wright. The Mendelian inheritance also had a numerical interpretation after incorporating the rules of heredity. Stephen Jay Gould and Niles Eldredge later came up with another theory that has a firm standing in evolution, punctuated equilibrium, by matching paleontological evidence with statistical evidence.[1] Andreas Wagner even introduced to us to a world of neighbourhood networks, using the mathematics of networks.[2] It is evident that mathematical concepts have been pivotal in understanding our world.

On the other hand, there are other acclaimed thinkers that had contributed significantly in their fields without the rigorous application of mathematics. In the same field of physics, the discoverer of electromagnetism, Michael Faraday, is known to have generated innumerable useful principles and concepts that shaped the understanding of the electromagnetic force. He, however, was known for not having as big a skill in mathematics. In medicine, Louis Pasteur, considered the James Clerk Maxwell of biology, formulated the germ theory of disease that is still used to date. Even in my field, medicine, scientists such as Rudolf Virchow developed insightful theories that initially had little grounding in mathematics. It dawned on me that theories were abstract nets that helps us fish information in a universal sea of data. While some were more mathematically inclined than others, they all helped us generate knowledge. Mathematical equations only intensified our understanding by encouraging high degrees of testability and thus falsifiability. The more specific the model, the higher the testability. This had been my problem ever since I started working on this theory.

In a field such as the one I pursue, medicine and surgery, there is hardly ever much that is stressed regarding mathematics. Some concepts that require algebraic applications are only noticeable to the keen eye. Calculus, which is the language that does not respect academic boundaries, is never touched, nor even nudged by my lecturers. While I gradually took steps to advance in mathematics, I took solace in some simple concepts along the way. The very first being the calculus of probability. If the probability of annihilation of a particle is P, then the tendency to avoid annihilation is (1-P). I loved its simplicity. However, before a mathematical equation is used to describe reality, a set of assumptions is vital in order to limit the domain of its application. The primary assumption that OS makes is that when two particulate organisms join, the overall probability of annihilation reduces only when the two particles result in a new emergent organism. It only increases if the aim is complete annihilation of the emergent *and* constituent organisms. I have been able to elaborate this in the previous chapters. What then are the consequences of such an interpretation? The first is the emergence of diversity.

Statistics: The Inverse Law of Large Numbers

After the discovery of the gene and Mendelian inheritance, Natural Selection underwent a 'mutation' sparking an interest from a narrated to a statistical interpretation of species and populations. Evaluating the impact of a gene the moment it appears in a gene pool can be very inaccurate. There is a wide latitude of interpretations that a newbie gene can get from scientific analysis the moment it is recognised. Novel genes are always being identified in various species, from ants to bacteria, but whose roles we know little about. Since conventional knowledge follows from what we have always known in the past, especially from fundamental concepts, the role of a new entrant, such as a novel gene, becomes rather difficult to immediately interpret. Geneticists would, however, likely correlate the new entrant to an evolutionary step, especially if it can be passed down in its stable state to various generations. Thus, before its effectiveness or lack thereof is established, its impact is very variable. It is not known whether it is beneficial, neutral or lethal. One of the core processes through which these novel genes emerge is through mutations. The process by which this mutation leads to these three outcomes is also not quite understood at this point, especially when the gene is relatively new in the gene pool. Therefore, statistically speaking, it is highly variable.

The high variability can be attributed to a mathematical law known as the law of large numbers. In a very broad sense, the law of large numbers states that there is more certainty to be gained by considering many variables compared to a few of them. You can know more about a group of people by studying a large portion of them, say 100,000 compared to a smaller proportion, say 50. It is this law that governs research in the medicine and other academic fields. In a poker game, you can tell the pro from the novice after a series of games, but not from a single game. In fact, it is the law that explains beginner's luck. You can tell much about a gene after a long period of time, compared to a shorter time frame, and much less at its point of emergence. This school of thought explains why genetic studies consider the *E. coli* or the fruit fly as an excellent specimen for experimentation. The *E. coli*, for instance, has a doubling rate of as low as 20 minutes in a laboratory, such that in a day, one can collect up to 72 generations.[3] To put that into perspective, in two days, the replication process will result in 2^{144} such bacteria, much more than the total number of humans ever lived! Such a method is warranted, since it focuses on the fittest of the genes from the gene pool. That is, the genes that have survived over a long period of time to be passed down from one generation to another.

Over a long period of time, we can say that we start getting signals about the gene's capability. In short period of time, we only get noise, which is hard to decipher. Very little to nothing can be learnt at this point. This follows from the *inverse* law of large numbers. If it takes a long time to be *certain* about the knowledge of something, that is, signals, then over a short period of time, we are *uncertain*; we get a lot of noise. This is not so much a law as a consequence derived from the law of large numbers. For a short time, there will be a lot of variation, but over a long time, stability.

Organismal Selection focuses more on the short-term, by considering that a particle is an organism. The first particle had to emerge for there to even be an archive that we could call the RNA or DNA (another particle and hence, another organism, for that matter). Modern synthesis teaches us that the gene can only express itself through the 'interactor interface', the organism. The gene is further *only* appreciated because of its expression, that is, the phenotypic manifestation. A gene for tallness is expected to express tallness in the organism. Knock-out the gene and you are expected to knock-out the expression. If I take away the glutamate at chromosome 6 of a human being, and replace it with valine, I end up with an organism with sickle-celled red blood cells. However, as Gould insists, without an interface for the gene to express itself, the gene is considered useless. Even more is that if the gene is to stay for a long enough time to merit consideration by Natural Selection, it must have an interface to account for its mettle. In this sense, the organism is vital. With some caveats, this is where the argument about the emergence of genes through the modern synthesis arguably ends.

Organismal Selection, on the other hand, has already shown that the organism in its simplest form, can result in the emergence of another through merging with another organism. A particle can merge with another resulting in another emergent organism. The types of forces that define the merger are varied, as I shall elaborate later. However, the two organisms are inextricably linked. The line of thought of modern synthesis has taught us two things: firstly, without the genome, there is no organism and secondly, that the organism facilitates the further survival of a gene. In this sense, the organism and the gene are inextricably linked. This is one instance that particles are linked, or as OS puts it, merged, via different types of forces.

What such a relationship portrays is that there is a straight line between the genotype and the phenotype – the genotype representing the genetic makeup of the organism and the phenotype being the physical manifestation of the genotype. The genotypic toolkit that an organism presents with at birth should therefore be the same genotype they should be buried with. Unfortunately, this is not the case. For instance, studies have shown that there are varying chromosomal pictures in different organs among dead specimen.[4] This is the curveball that the building body of knowledge has thrown at Natural Selection.

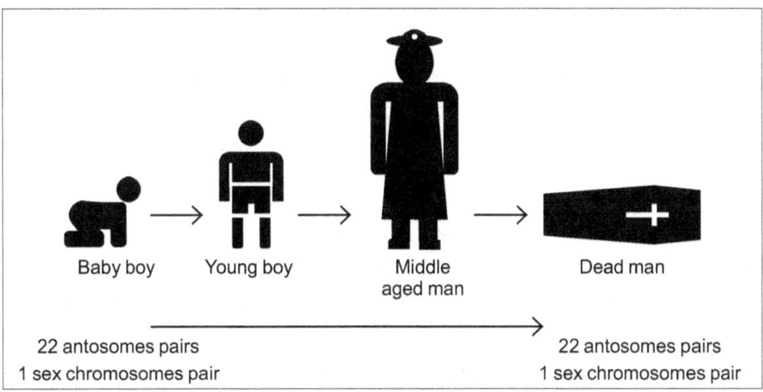

The hypothesis of the preserved genotype from birth to death. This is no longer corroborated by studies that show different genotypes across the lifetime of an individual.

Natural Selection holds that life evolves through a series of steps from the already available genes. That is, evolution should happen through a common genetic toolkit. The combinations of these genes provide the limit with which an organism is bound to evolve. Humans have around 21,000 genes, creating an astronomically large number of combinations. Modern synthesis states that evolution should use the combination of these genes as the playground for evolution of the organism. In essence, the genotype you had the moment the sperm fused with the ovum to form you, the reader, should be the same genotype you should have when you die. Seirian Sumner has elaborated how advanced technology has enabled us to sequence these genotypes *de novo*.[4] The data available for such *de novo* synthesis is present in large volumes for the algae, the leech, the butterfly, the dolphin and even humans. Furthermore, in each genome there is an endless series of unique genes that have been identified. Sumner notes that approximately 20% of genes in nematodes are unique. Each lineage of ants contains about 4000 novel genes, but only 64 of these are conserved across all seven ant genomes sequenced so far. That is a startling 0.016% that has been conserved. The rest are novel genes. These novel genes are proving instrumental in the evolution of biological innovations in the freshwater polyps, the worker caste of bees, the wasps and ants and even the regenerative potential of newts. In humans, they are even associated with diseases such as leukaemia and Alzheimer's.[5] The novel genes are *not* from a conserved genetic toolkit. They are novel in every sense of the word. Where, then, did they come from? Before we answer this question, there is more that shakes the idea of a common genetic toolkit. It has to do with mosaicism.

Mosaicism is the presentation of two or more cell populations with different genotypes in an individual.[5,6] To acknowledge the presence of mosaicism one needs to get the karyotype, that is, the picture of a genotype. This is just what was collected from numerous individual post-mortem reports showing varying degrees of mosaicism. The genotype of the liver was variably different from that of the brain, and that of the brain from the stomach. Different organs displayed a different genotype. Mosaicism has also been seen in both healthy individuals and those with different diseases.[7,8] This is the other curveball that has sparked a lot of questions and even cast doubts at the foundation of the school of thought on the evolution of an unchanging genotype. It appears that the development of the organism and its evolution does not happen from a particular genetic toolkit.

These reports highlight the one pitfall in the long-term view of evolution. By focusing on the genes that are able to remain strongly competitive and stable enough to last for many generations, these small bumps get overlooked. Fortunately, this is the primary focus of OS – the short term. This narrow time frame was the first evidence I used to conclude that the organism was the fundamental unit of diversity. The inverse law of small numbers asserts that, in short periods, there is bound to be significant variation. The previous two paragraphs show just that. In the short lifespan of an individual, there is bound to be untold mosaicism. Using evidence formulated by genetic studies, OS has demonstrated, if not predicted, how diversity lies fundamentally with the organism. Since the DNA also is an organism in the story of OS, mutations support the idea of the organism being the fundamental unit of diversity. It introduces the first of the four S's that will help in understanding the arrival of the fittest – statistics.

Statistics owes a significant chunk of its development from the law of large numbers. It is the basis for the use of large data sets as sources of developing trends or highly reliable algorithms, such as the kind that Google uses. In fact, a lot of the formal graphs and formulas in statistics bear in mind an infinite set of variables. Infinity is the ultimate large number. On the other hand, the inverse law of large numbers, popularly known as the law of small numbers, highlights that the most extreme of outcomes come from small samples. Daniel Khaneman and Amos Tverseky are known for clarifying the marked human insensitivity to sample size and its profound consequential effects. In fact, the law of small numbers illustrates how there is no need for a causal link to explain an extreme outcome from a small sample size. The extremely well-performing schools are likely to have small numbers, just as the poorly performing ones. If small samples are notorious for extreme events, what of the *smallest of the smallest* sample sizes? Say, a solitary organism? Anything less than an organism is not an organism however you would want to define it. Thus, the solitary organism is the smallest sample size one can get, and with that, it should have the most extreme outcome. Thus, using this formal argument, the organism should be the fundamental unit of diversity. As small sample sizes are notorious for extreme events or traits, the smallest of sample sizes should have the most extreme event. This is the source of variation among organisms. No causal basis, just mere existence without yielding to the genetic errors of DNA replication, mutation or sexual reproduction. In fact, what OS says is that by these processes the new DNA mutation constitutes a new organism, diverse in its own right by mere existence. A particle, just by its mere existence, accounts for variability. This is in agreement with the map, which is just as robust now as it was when I had introduced it – the map of an organism and its universe.

What are the implications of such an interpretation? Firstly, the very first organism is therefore a victim, if not the proud winner, of untold variation. The very first particle, the first cell or even the first multicellular organism bears an inherent, untold variation that demands no causal link, just mere existence. Only a quantum computer can begin to accurately predict the possible outcomes for an organism. What about the common ancestor? A thought experiment might help.

Natural Selection speaks of a common ancestor who, perchance, was successful enough to reproduce. However, had the ancestor not succeeded in this, Natural Selection would not have conceived a mechanism of variation. According to Natural Selection, variation happens through genetic mutations and re-assortment. This interpretation, nevertheless, is not false. It just highlights a narrow scope of variation. Every time an error or mutation in the DNA occurs, we have a different compound. This is also a form of variation because a mutation creates a different organism. But we need not succumb to the errors of replication. We need only yield to existence. Existence can even be appreciated in the absence of the nucleic acids. It can therefore be extrapolated to other particulate organisms. This interpretation destroys the idea of there being a significant leap between particles and the first RNA molecule. By the mere existence of particles, there have been organisms all along even before the RNA world. The inverse law of large numbers, thus, when stretched to its limit, attaches an unscalable value of variation to an organism, whether particulate, cellular or multicellular.

Secondly, the first organism becomes an essential component of evolution. From Natural Selection, it appears that the common ancestor was able to reproduce. We cannot, however, tell if this was successful on the first try, or even on the second. Our current understanding of how evolution works shows that it should not have been the case. There ought to have been several tries before reproduction, as we understand it, should have happened. Even then, the first ancestor that tried and failed to reproduce existed. Just because they existed, they displayed untold variation. It is untold because even the very first organism that managed to reproduce could not be predicted. As OS paints it, the cosmological short-term view, is at least as important as the long-term view. A simpler way of stating it is that there would never have been a long-term view if there was no short-term view.

Organismal Selection does much more than focus on the long-term view. It highlights that there are organisms that had existed longer than genes. Genes that can last long enough will have odds in their favour if they are short and stable. A short one will be easy to duplicate. A stable one allows it to maintain its state for a long time over generations. However, there are other organisms that have lasted just as long if not longer than genes even though they have never reproduced in the biological sense. Quarks, protons, neutrons, electrons, atoms, molecules, all these existed and continue to exist long after genes have been annihilated. So even if we are to consider the long-term view, particles still have warranted consideration in evolution. The short-term view opens a different perspective about the long-term view!

At this point, it should be clear that OS differs from Natural Selection in the sense that, when particles merge, new organisms are registered, without specifically talking about reproduction. Granted, particles merging does imply some form of conjugation, much like the process of fertilization, which evolutionarily speaking, should be the hallmark of sexual reproduction. But does it then mean that OS predicts Natural Selection? I do not think it does, and if it does, then, not directly. I think that it does, however, explains Natural Selection much better – in the short term, we have a lot of noise, but in the long term we have signals. In the short term there is a lot of instability but in the long term, the stable genes are evident. In the short term, there is untold variation, but in the long run, a modicum of certainty. It even explains it better by considering a counterfactual element, better appreciated by the following thought experiment.

If by some chance, there appears a human being who is able to achieve self-replication, not the typical one marred with errors, but perfect self-replication, the gene-centred view ascertains that there is no variation between these two people. However, OS predicts differently, insisting that they are different, due to mere existence. This not only shows how different the two theories are, but indicates the limits of Natural Selection; its limitations are identified by the errors that come with replication, which is attributes to variation. Ironically, these errors are supposed to amount to a gene with better survival potential. This is hailed as the mechanism that makes Natural Selection possible. It is one of the principal ways that OS differs from it. The emergence of an organism happens when a particle appears or when a particle merges with another. It happens when a cell merges with another. This further explains the arrival of the fittest in the fitness landscape through the second S – symbiogenesis.

Symbiosis and Symbiogenesis

Symbiogenesis or serial endosymbiotic theory is the leading evolutionary theory that explains the emergence of eukaryotic cells from prokaryotic organisms.[9] The evidence in support of this theory firstly lies in the mitochondria. The mitochondrion is the power-house of the cell, generating energy for the eukaryotic cell. The principal difference between a eukaryotic cell and a prokaryotic cell is the internal cellular constituents, known as organelles. It has classically been known that in a eukaryote, the organelles are membrane-bound, but in a prokaryote, they are not (presently, there are various organelles that have been discovered to be membrane-bound in prokaryotes[10]). In a eukaryote, the mitochondrion is one of the various cellular organelles. It however differs from the other cellular organelles in the sense that it has its own DNA floating freely like that of any known bacteria. They are not bound into chromosomes neither are they covered by histone proteins (a group of proteins typically binding DNA in the eukaryotic cell). Additionally, mitochondrial ribosomes and those of respiring bacteria are sensitive to the same group of antibiotics such as streptomycin. There is no other theory that explains this mystery other than symbiogenesis, which states that two unicellular organisms merged. Numerous studies have been able to trace the mitochondrion back to Rickettsiales group of bacteria.[11, 12, 13] Therefore, in eukaryotes, mitochondrion forms independently of the nuclear DNA.

Similar arguments can be used to explain chloroplasts, which are organelles responsible for converting sunlight into potential food for plants.[12] Chloroplasts and mitochondria contain transport proteins on their membrane that bacteria have on their cell membranes. Specifically, chloroplasts can be genetically traced to the category of bacteria known as cyanobacteria. So far, symbiogenesis is the only theory that adeptly captures this phenomenon. This hypothesis was first described by Konstantin Mereschkowski and articulated and supported by the works of Lynn Margulis.[13] Symbiogenesis has since then grown to be appreciated in nematodes, amoeba, plants and animals. Where did this strange idea come from? The truth is, we can only guess. However, what I propound in this book is almost in every bit as strange as the idea when it was first proposed by Mereschkowski.

I have already shown how a new organism emerges when a particle and another particle merges. The sodium particle can merge with the chlorine particle to form a new organism, conveniently called sodium chloride. This behaviour is universal, where particles seek to merge with others, to avoid annihilation following the calculus of probability. Seeing as we have classified particles as organisms, it should not strike us as odd that one prokaryote would merge with another. In short, there was *particulate* symbiogenesis even before bacteria existed. This happens between two separate organisms, which would merge, form an emergent organism, and consequently, its new universe. This is how the second 'S' explains the arrival of the fittest.

In particular, symbiogenesis is appreciated at the genomic level. What this means is that there are genomes that can be traced to the constituent organisms before the merger that formed through symbiogenesis. However, before symbiogenesis was appreciated, there was symbiosis. Symbiosis is the relationship between two or more organisms where one benefits the other. A symbiotic relationship can lead to symbiogenesis, but not always. Usually, this merger is special in the sense that one organism ends up benefitting the other and vice versa, without the other being aware of it. This is a keystone feature of what is considered in system dynamics as hierarchies. In hierarchies, the role of a subsystem serves the other subsystem and vice versa. In the case of organisms, by one seeking to avoid annihilation, it serves another, and the same works for the other organism. In this sense, the zero-sum game does not quite dominate since the roles of the individual organisms ends up serving all the involved organisms. For instance, the termite can harvest wood and break it down into pulp but cannot digest it. In the termite gut, however, there exist microbes that can digest wood pulp but cannot harvest it. And there is more. In the same termite mound, the waste they produce after incomplete microbial digestion ends up serving fungi who need the dead material. The fungi then grow in size and get pruned (consumed) by the termites to form tough, swollen shoot tips that prevent the fungi from reproducing. The fungi, however, continue to grow in size.[14] The whole termite mound ends up acting like an entire self-regulating organism, with no central control. It grows just like any other biological organism, without any idea of a nuclear material. The role of one organism ends up serving the other, forming a stable hierarchical relationship. There are numerous other examples, even at the family level in our homes.

Symbiosis and symbiogenesis further highlight an important perspective that I have delayed in talking about – the characteristics of mergers. From the interaction between sodium and chlorine to that between a prokaryote and an endosymbiont, it becomes clear that merging is an appropriate term due to the differences between these two examples. Sodium and chlorine will merge by forming an ionic bond. That of the endosymbiont and the prokaryote merges, but not in this sense. There is, however, a merger, as evidenced by the numerous observations in support of the symbiogenetic theory.[14] The general sense of the term merger extends past the steric and electrostatic complementarity of particles. In eukaryotes, another organism had to be engulfed completely for it to be inside another. An atom can also do something similar, by taking another organism, an electron, resulting in the formation of a new organism – an ion. But if another atom were to merge with another, it would be based on other forms of mergers, such as steric or electrostatic mergers. The different types of bonds described in chemistry constitute the structurally appreciated mergers in that field. And with each unique merger, stems a unique property in the emergent organism. Emergent properties then typify emergent organisms. Taste, for instance, is present in the $C_6H_{12}O_6$ compound of glucose, but it is independently not present in carbon, hydrogen or oxygen. The term merge exemplifies a dynamic relationship between particles, the kind that results in emergent organisms with emergent properties. It is not static as the term 'electrostatic' connotes. I will further explain how extensive this dynamism extends into the process of reproduction later in chapter seven.

What this merging concept further highlights is the constraints of scientific analysis, which tends to break down structures into their primary elements for study. Such studies have pioneered the discovery of very small particles with a lot of practical use, such as Feynman's contribution to quantum electrodynamics. This method is a bottom-up approach. Organismal Selection shows how the achievements of such a system have been so luminous that it often blinds one to such possibilities as the kind that OS perpetrates. Incorporating both bottom-up and top-down perspectives gives us a better understanding of our shared reality. Notice that I say 'shared reality' since the map of the organism and its universe still stands. I shall elaborate much more when discussing language and how it is also related to mergers.

This bold strategy of merging with resultant emergent properties is so robust that it holds even at the subatomic level. One of the simplest particles known to man are quarks. Significant tests have been carried out to affirm their existence. For instance, quarks have a property known as spin. Spin is also observed in other particles, such as electrons. Wolfgang Pauli, an Austrian scientist, is famous for his exclusion principle, which states that no two electrons as an electron pair can have the same spin.[15] This property can easily be interpreted as either clockwise or anticlockwise – as opposites. Spins can also be described as either half spin or full spin, depending on the particle. This principle holds even at the sub-atomic level to include quarks. Thus, it was assumed that the spin of a single quark should able to contribute to the spin of the stable tripod quark. Independent experiments have shown that this is not the case. The spin of one quark depended on the other two. Hence, the overall spin of the three quarks cannot be predicted from the individual spins but only after the quarks have merged. This is an excellent demonstration of emergence – the overall 'tripod spin' is emergent from the three quarks that merged to become a single entity. The tripod displays features of a single organism.

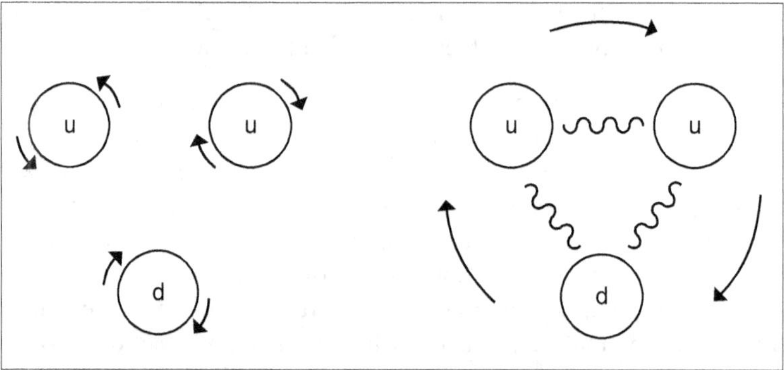

Quarks and spins. The spin of a stable 'tripod' quark system cannot be predicted from the individual spins of the quarks when separated.

So far, two or more particles merging and an endosymbiont and prokaryote merging result in the formation of a new organism. It is then safe to conclude that there is a *similarity* in the bold strategy of merging across different scales. There is also a *similarity* in that there is an emergence of a new organism. There appears to be a property of self-similarity across descending scales of an organism – the eukaryote, the prokaryote and particle. Self-similar patterns in nature are known as fractals. With organisms seeking mergers, there appears to be a fractal pattern of *behaviour*.

Fractal Geometry of Nature vs Fractal Behaviour of Nature

The idea behind fractal geometry was coined by the Polish-born mathematician, Benoit Mandelbrot.[16] The word fractal means roughness or ruggedness, a trait of nature that interested Mandelbrot. It is this interest that made him discover that in nature, there are elements of self-similarity in objects that aggregate to form a rough superficial appearance. He understood, as many that preceded him, that the perfect symmetry of Euclidian geometry could not explain the ruggedness of natural features. They appeared fractured, hence the term fractal. An example of fractal geometry is seen in the cauliflower, which appears the same as the complete vegetable, even when scaled down. It is as if there are mini-cauliflowers that aggregate to form the global cauliflower. If two images are taken, one of the scaled-down version of the cauliflower, the mini-cauliflowers, and another of the global cauliflower, they appear so similar it is striking. It is from self-similarity of geometry that the roughness of nature is better appreciated.

An up-close image of a cauliflower. The fractal geometrical features show the self-similarity at different scales.

A simple pattern such as a repeated self-similar pattern, accounts for a lot of complexity seen in nature, such as the branching of capillaries and bronchioles in the human body, clouds, mountains and even coastlines. Numerous computer algorithms have incorporated various fractal models which later became adopted by several blockbusters. The

Lord of the Rings used fractal geometry to create images that boarder reality as much as believably possible. It all followed from Mandelbrot's ability to formalize that the fractal geometry of nature was the appropriate branch of mathematics that dealt with the ruggedness of objects and physical features. While his interest was on the *self-similarity of structures* at different scales, I would like to focus on the *self-similarity of behaviour* at different scales of an organism – particulate, multi-particulate, unicellular or multicellular. I tend to think that the story of the magnet resonates with this idea of a fractal pattern of behaviour.

After Michael Faraday discovered some of the laws of magnetism, he was immediately propelled into scientific fame. He, however, mentioned some issues that he could not quite wrap his head around, the first being the magnetic lines of force. He was puzzled by how they were so discretely aligned, never crossed each other and how they all emerged or converged in the poles. The second one was the one that I considered mystifying – the polarity mystery. Firstly, the poles are arbitrary. There is no *a priori* universal law that shows that this will always be the North or South Pole before testing it out with some guiding rules or with another magnet. The second was that if you split the magnet into half, the same structure of the North Pole and the South Pole remains for the split magnet. The poles remain arbitrary points that are not fixed even after splitting the magnet. There is no split magnet that only has a North Pole or the South Pole alone. They both have two poles. The behaviour is self-similar across descending scales of the magnet. This is generally a rough example of what I call a fractal pattern of behaviour.

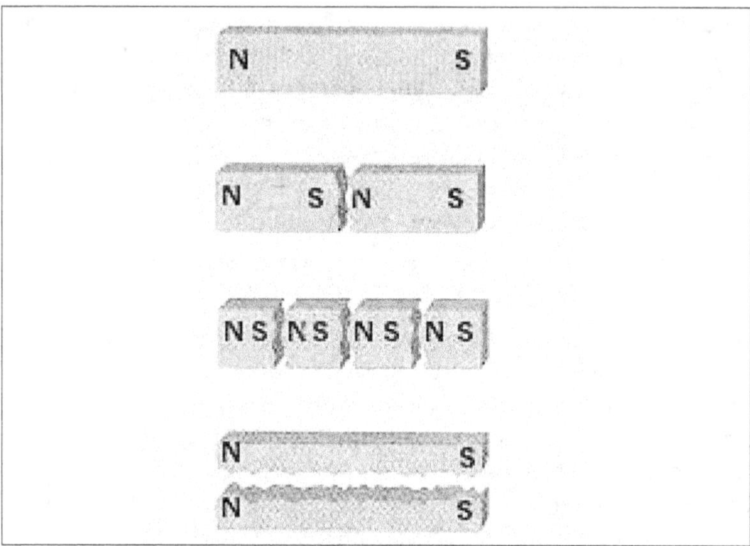

Magnets split from larger to smaller scales. By preserving their polarity, magnets also display a fractal behavioural feature at different scales.

The first hint that shows a self-similar pattern of behaviour in the organism has to do with the tendency to avoid annihilation. At different scales of an organism, from the star to the blue whale, the elephant, the cow, the rat, the shrew, the atom, the electron and the quark, all tend to avoid annihilation. In this regard, they are all scaled versions of each other. There is also another elusive one that I find intriguing, which I extracted from the works of Daniel Kahneman.

In his book, *Thinking, Fast and Slow*, Kahneman describes, using a model known as the fourfold preference model, that people were not always risk-averse, but that there were instances where they could be risk-seeking, as shown in the table below.

	GAINS	LOSSES
High Probability	95% chance of winning 100,000 Fears the slim chance of losing Accepts the unfavourable settlement **Risk-averse** (The Rich/Wealthy)	95% chance of losing 100,000 There is a slim chance of winning Rejects the favourable settlement **Risk-seeking** (Gamblers)
Low Probability	5% chance of winning 100,000 Hopes to bag a huge win Rejects the favourable settlement **Risk-seeking** (Lottery-Ticket buyers)	5% chance of losing 100,000 Fears a large loss Accepts the unfavourable outcome **Risk-averse** (Insurance seekers)

In an anticlockwise manner, the top left includes those who are rich enough to settle for something rather than nothing. They have a high probability of winning but the slight chance of losing is too disappointing. For them, a bird in hand is worth two in the bush. They settle for less, and accept the unfavourable outcome. They are risk-averse. The bottom left comprises those that have a low probability of winning but still end up trying. Those that buy lottery tickets have a very slim chance of winning, but they bank their hopes on this slim chance, and are hence, regarded as risk-seeking. The bottom right corner exemplifies those that seek insurance, especially after a significant calamity. Often, the calamity has a small probability of occurrence, but for peace of mind, prefer seeking insurance than risk the slim chance of you or your family being a victim. Thus, they settle for the unfavourable option. These individuals are risk-averse. The last box is the one that I loved the most – those who evaluate between a low probability of a big win and a high probability of a loss. These individuals fear the prospect of a near-certain loss, and consider the slim chance of taking home the big win. Movies depict this exceptionally by the last die toss that the man has, who at times asks the pretty lady by his side to blow into it, hoping it will land him the big win. These individuals are risk-seeking. This is the one trait that is seen across all scales of organisms. How?

Like the red-shifting of distant galaxies, the probability of annihilation is always shifting towards one, towards certain annihilation. Organisms can either decide to settle for this near-certain loss or seek the slim chance of the big win. That an organism has a universe inclined to annihilate it, this slim chance of a big win is one that is an appealing gamble for the organism. They would therefore traverse their universe for another organism in the hope that they would merge and get the big win; postponing their respective annihilation. At the subatomic level, this is seen by the quark. A solitary quark seeks mergers with two others for it to be stable. They then demonstrate asymptotic behaviour; that is, the particles behaving in a manner as if resisting separation. The farther the separation, the greater the resistance. Atoms react with other atoms to create compounds. A prokaryote would engulf an endosymbiont to avoid near certain annihilation. There would be sexual intercourse for there to be a merger, initially between two individuals and ultimately between two gametes. The big win is this – an emergent organism and delayed annihilation. Evolutionarily speaking, organisms are gamblers. They are risk seekers. The fractal behaviour of organisms extends across all known scales. Just as fractal *structure* of nature introduces some aspect of complexity, so too does the fractal *behaviour* of nature. Yet again, we encounter how OS is not just a theory of diversity but also of complexity. The rule of the game remains – anything to avoid annihilation. As an exception, there is more to sexual reproduction as a merger than the simple merging of gametes. I will dwell on that later. There is however one aspect that I want to tackle at this point, as a consequence of mergers, and that has to do with variation.

I am yet to come across a book or article that comes close to adequately explaining why an organism, as complex as those we have been able to scientifically study so far, came into existence. There is only marvel at such complexity and few attempts at giving explanations. For instance, the concentration of particles in the human body borders astronomical levels. Just counting cells, humans have an average of 10^{12} to 10^{16} cells. This is only cells. It is estimated that in the known observable universe, there are around 10^{89} particles. The universe referred to here is the kind that astrophysicists have been studying for years. The ranges might appear far off, but one wonders why a large concentration should be in a single organism such as you and me. Organismal Selection explains in part that this is possible, again, using the law of small numbers.

To build on this argument is this thought experiment. Assume a room is filled with ten balls. These balls have the ability to fuse with each other. The room is analogous to the universe and the balls to the very first family of particulate organisms. The balls are the 'originals'. Each ball is an organism and anything external to it is its universe. From the law of small numbers, the variability and hence diversity potential for a single ball is significantly immense; it has unscaled diversity. However, from the perspective of *all* the available organisms, the different diversity potentials formally cancel out and the diversity of the shared universe (room) becomes relatively low. One ball then decides to fuse with another forming a new, emergent organism. The degree of diversity in the shared universe (room) immediately increases. The sample size has reduced, and the diversity potential of the shared universe (room) increased. If more balls fuse, the sample size again reduces but the diversity potential of the universe increases. The unscaled diversity increases in line with the law of small numbers. More mergers reduce the sample size and thus increase the level of diversity among the available organisms. The keen eye will notice that this thought experiment couples statistics and symbiogenesis, to display how diversity and complexity are linked.

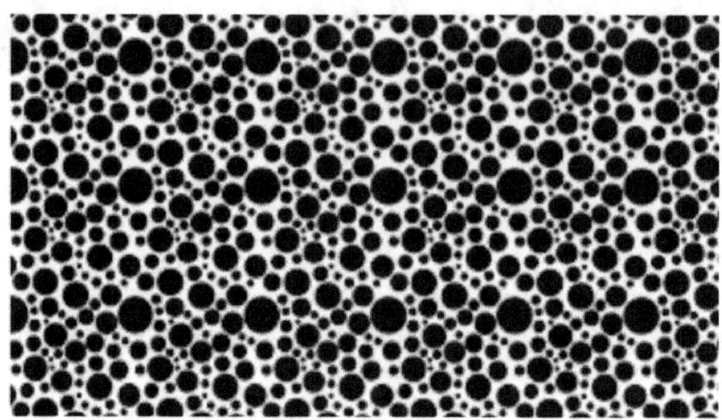

A simplified two-dimensional version of a room filled with spherical balls. There are smaller balls than larger ones throughout the room. By capturing statistics and symbiogenesis, the experiment illustrates the relationship between diversity and complexity.

The thought experiment further explains the emergence of life as we have always known it. When more particles merged to form more complex molecules, such as the precursors of the RNA and DNA, the diversity potential of the universe was much higher than when there were single elementary particles. Life with reproductive potential emerged as a property springing from the overall increased diversity of the universe once particles started merging and with merging comes complexity. The one highly plausible way this happens is if the merged particles form a new and different entity, thus minimising the sample size while simultaneously increasing the diversity potential of the universe. It is such explorations, stretching the limits of OS that were mind-blowing, so much so that I had to share it with the world, to, if not appreciate its beauty, critique and drop it at worst or at best, develop it.

Even more astounding is how OS could explain the Cambrian explosion, which took place around 540 million years ago. Two billion years ago, oxygen was deposited into the iron sink, forming ferric oxides. Since these compounds were insoluble, they settled on the oceans and sea floors. As a result, the oxygen rates in the atmosphere were around 1-2%. The dominant organisms were prokaryotes specialised in anaerobic forms of respiration (metabolic processes that do not demand oxygen).

By 1.8 billion years, the first eukaryotic organisms appeared, alga-like forms and oxygen levels began to rise as these organisms produced oxygen through the mainstay process of photosynthesis. It is believed that near the end of the Proterozoic era, the first animals evolved. Around 540 million years ago, near the beginning of the Cambrian period, a pivotal event happened in the evolutionary tree – animals increased in size and diversified explosively. Oxygen level at this time was nearly 21%, the same level as we know it today.

While competing theories have centred arguments on the relationship between the Cambrian explosion and the rate of oxygen levels, OS can offer an alternative explanation. Using the same thought experiment of balls inside a room, the diversity potential in a universe is seen to increase with the reduction in the sample size in that universe. The reduction in sample size is matched with an increase in size (and complexity), which can only happen through mergers. This simple thought experiment captures in a very simplified way, the Cambrian explosion, the so-called Big Bang of animal evolution. The pace of the diversification could be attributed to the diversity potential present at the time, which was increased by particles and organisms merging. As I shall also explain in Chapter 6, size ended up being useful in preventing organisms from being destroyed. The double reward of mergers (symbiogenesis) and size would be adequate to explain the Cambrian explosion.

Furthermore, the concept of merging goes ahead to explain how a particle as large as a star, can merge with that as small as a hydrogen atom. The star is a large organism, merging with another small organism, the atom (stars can also merge with one another). The cell in the liver can merge with a smaller particle generating a different genotypic picture from that of the brain. This could explain the different karyotypes (genotypic picture of cells) have previously been reported and the mosaicism in both healthy and disease-bearing individuals. To top it all up, there is a tempting incentive to merge. The fourfold pattern of preferences has shown why this strategy is tempting. The gamble has high returns even though the probability of getting this big win is small! It is much more appealing than near-certain annihilation. Organisms have much to lose but much more to gain!

But first, let us consider for a minute the diversity of a single organism. The law of small numbers tells us that *extreme* outcomes are expected in small samples. This is a statistical outcome not amenable to a causal link. Organismal Selection stretches this law to its limit, using it to confidently assert that the fundamental unit of diversity is the organism. It makes it all the more interesting when considering the very first organism, either particulate or as we understand them biologically. If we take the virus, for instance, they are simple organisms. Much simpler than bacteria. Following our baseline definition of a particle as an organism, a virus falls under the same category – as an organism. The basic and ubiquitous function of a virus is reproduction, even though it needs a cell to achieve this. We can then use reproduction as a feature for the basic living organism. But this is a trait that is seen in virtually all organisms. It is not extreme. It is nonetheless extreme when we consider the very first organism that thought up the strategy of reproduction. That must have been extreme; extremely bold if not diverse. For it to be a stable strategy, it must have braved all other strategies, to the point that it is used by all the living organisms as we have always known them. Reproduction, therefore, tells the story of the arrival of the fittest. This is different from survival of the fittest which is the *modus operandi* for the modern synthesis. Ironically, reproduction, the one process that raises the legitimacy of Natural Selection, appears to explain the arrival of the fittest, but emphasis has often, if not always, been on the survival of the fittest.

The modern synthesis, which I consider a 'standard model' in the field of evolution, is very good at explaining the survival of the fittest. In fact, it uses a model that mathematicians and myself (since I am not a certified one) like using – inversion. Inversion of a problem helps in finding a solution by working backwards. Once *Staphylococcus aureus* bacteria develop a means of resisting methicillin, through inversion, the modern synthesis explains how those that are currently existing survived. The existing ones 'fit' the prevailing conditions. This is what 'fittest' means.

However, it is not proficient in explaining the arrival of the fittest partly because it does not take into account the rich spectrum of diversity that is attributable to the organism. It only explains what exists, how it has been successful and how it fits in the fitness landscape. But it does not explain very well, as it does survival, how it arrived in the landscape in the first place. Organismal Selection inverts this process and explains the arrival of the fittest, so far, through the first two S's – statistics and symbiogenesis.

The inverse law of small numbers further explains Darwin's quandary. Darwin marvelled at how there was a rich diversity of organisms when survival of the fittest should have minimised his diversity.[17] Freeman Dyson called it the 'diversity paradox'. However, Motoo Kimura, in his theory of neutralism, shows how diversity can stem from small populations rather than big ones.[18] Organismal Selection is in agreement with Kimura's theorem, but stretches it further by emphasizing that: the fundamental unit of diversity is the organism. Undisputedly, gene-centred view of evolution states that the fundamental unit of inheritance is the gene. I do not deny this (even though there are other forms of inheritance besides the genetic type). What can easily be overlooked, however, is that when the population increases, the overall diversity starts cancels out, and we are left with similarities, or even a homology, and that is where the idea of the most fitting gene emerges. Organismal Selection, therefore, agrees with the long-term view, that the gene is a testable and reliable unit of inheritance, *only* in a shared universe. It does not stop there. It further stresses the relevance of diversity that is not just gene-centred, but organismal-centred. This is tied to the definition of evolution we described earlier – the ability to avoid annihilation, a property seen in both the organism and the gene. The gene is found within the organism and is therefore included in the entire sum or product of what an organism constitutes. It is a manifestation of the unscaled diversity seen in an organism. And it further insists that my universe is not your universe. It only appears so, but it is not shared. Thus, while the OS theory insists on a *unique universe* that an individual organism brings forth, the gene-centred theory insists on a *shared universe.*

This explanation helps us understand why evolution is best studied in islands since they give a powerful mix of diversity stemming from relatively small samples in the island, and inheritance, from the idea of a shared universe (i.e. a shared island). Darwin even gathered a good portion of his data while in the Galapagos. An island is no different from a petri dish, where experiments on evolution can be conducted. The samples in this 'island' of a petri dish or even in an actual island, are usually relatively small. The reason why emergent traits and organisms wielding them are noticeable can be attributed to the small sample size.[19] Diversity is therefore rich in an island compared to a large continent. When Jared Diamond set out on a bird expedition in the Papua New Guinea islands, he describes how he encountered a lot of variety. What about the earth? The earth is an island relative to a mass as extensive as the Milky Way. With regard to the species that occupy the Earth, we have barely scratched the surface. Even then, over 90% of species that have ever existed have gone extinct. The diversity can again be attributed to the small sample size, in line with the inverse law of large numbers, a central pillar of OS. Darwin's diversity paradox, thus, dissolves.

One theoretical model that I believe corroborates OS is that of dissipative structures. The concept of dissipative structures sparked a lot of interests in physics and chemistry, especially since it displayed structures that evolved contrary to expected logic at a time when thermodynamics, as a field, was fairly well established. Ilya Prigogine, the Nobel Laureate, along with other prominent scientists at the time, such as Lord Rayleigh, studied and confirmed that heat, which should be dissipative, could create ordered structures.[20] Heat was and is capable of creating discrete structures that display a remarkable degree of dynamic stability – a state far from thermodynamic equilibrium (a form of annihilation). Since the state of these structures is far from equilibrium, they were given a convenient term – non-equilibrium states. The study of dissipative structures then developed into the study of system states using what is now regarded as non-linear mathematics.

When considered as organisms, dissipative structures can and do encounter particular points in their respective universes known as bifurcation points. At bifurcation points, some particles that formed a dissipative structure have to move to one of either sides. At these points, there is an irreducible random property that makes it impossible to predict the outcomes. It is much like a river that arrives at a fork, at the delta. One might have traced a particle with all the known details across its path, but at that fork, the bifurcation point, certainty dissolves and there is no knowing which side the particle will follow. This hints on diversity. As dissipative structures move far from equilibrium, they move from a fairly understood dynamic state to a highly unpredicted and unpredictable state past the bifurcation points. These points also describe the emergence of novelty and uniqueness. They describe the emergence of unscaled diversity. From the rather new form of nonlinear mathematics of stable non-equilibrium systems, we arrive at what we did by inverting the law of large numbers – unscaled diversity. What Prigogine teaches us is that certainty is lost firstly due to non-linearity. Non-linear equations usually have more than one solution. The more non-linear, the more the solutions. The other way certainty is lost is from the irreducible randomness at bifurcation points. The existence of bifurcation points abounding unique systems and the fact that they account for the irreducible outcomes of particles demonstrates how and why they are critical points that bring order from chaos. This is in line with OS; how an organism is particularly unique in its path, past, present and future, also agrees with the robust map of an organism and its universe. Non-equilibrium thermodynamics, dissipative structures, bifurcation points and degrees of uncertainty are features typically discussed in the topic of complexity. Yet again, OS shows that it does indeed warrant discussion in explaining complex structures.

Taking all that we have described into perspective, it paints a whole new picture about the emergence of life. We do not know what life is, only some of its features, which we consider consistent with our vague definitions of it. Let us, however, first appreciate some of its wonders among humans. The first human being, it is posited, hailed from Africa. As the first of its kind (as is anyone who is born) they are bound to wield untold diversity. From this first creature, nobody thought that he would develop a bald-spot let alone invent a machine that would take him to the moon or even in space. Nobody would have predicted that some would develop light skin as in the case of the Caucasian. It would never even have occurred that an albino would have emerged. We have seen the trait of albinism in humans, but it also exists in mice and rats. What are the odds that a hermaphrodite or pseudo-hermaphrodite would have existed? What of the human with sickle cell disease? And what of the tribes? And their different languages? That a human being would be living together with the very beast it once hunted? That he would discover farming? Heck, even the ants and termites do this!! Who would have thought that our vocabulary would be as rich as to contemplate thought experiments and articulate them? That the mandarin alphabet would have a richly diverse set of characters? That a single person would be responsible for the death of so many Jews? That a man imprisoned for over two decades in Robben Island could decide, against human tendencies, to forgive the very people who imprisoned him for that long? Nobody. That is the untold diversity that stems from a single organism. Be it with errors or with its merits. The organism is so diverse that we are unable to put a figure to it or charter its trajectory. In short, the organism contains unscaled diversity.

What of the organism as OS defines it? As a particle? Does it have the diverse attributes that we accord the biological organism? A simple answer to this can be derived from another question, whether the study of particle physics is still alive as a subject. There has never been a time when particle physics was as alive as during this era. Don't believe me? Just whisper 'Higgs boson' into a particle physicist's ear. Factually, there is no telling that the first particle (whoever it was) could have given us hydrogen and its properties. That the hydrogen particle could have variations of itself. That from this particle, the prediction about the existence of quarks was possible nor even the Big Bang or the Big Bounce. And even more astonishing is the existence of virtual particles that cause quantum perturbations in what we previously thought to be vacuums, devoid of energy. The story checks out. A particle can be considered an organism. More importantly, let us not forget the map, of an organism and its universe. As long as an organism has its own universe, it is never the same as another. It can only be identical, but not the same.

Where does that leave simple concepts such as diffusion and osmosis? Or even active transport? While science only explains *what* these processes are and *how* they unfold, OS explains *why* they may manifest as they do. It does this by sticking to the fractal behaviour of organisms. A bridge that links biological organisms to chemical elements is a process we have previously discussed called the Turing Mechanism. This mechanism talks of reaction-diffusion models. Reaction-diffusion is a form of diffusion. This will be handled in-depth in the latter stages of the book.

At this point, I cannot shake the feeling that you might still be going through a difficult time contemplating that a particle can even be considered as an organism. For the few or the many adamant, there is an idea that can persuade you to at least consider particles as organisms. In a city, when you walk into your office, if you have one, you can pick out the bee-hive of activities in the work place. When you leave the work place, you notice an even bigger web of activities among people and structures that they might not even be familiar with. For the eagle, who sees all this from above, it only sees big moving blobs. With its keen eye, it even spots the smaller ones, the ones that it hunts. What of the naïve pilot who sees it all in a more inanimate way? The pilot only notices small dots moving in what she may perceive as random motion. However, if you ask the person who walks out of the office with you where they are going, their answer does not yield any hint of randomness. If you ask the eagle, if it were able to respond or if you were we able to communicate with the eagle, it would reply that the small prey that they hunt are trying to make it to their destinations without turning into supper (after which the eagle would give a diabolical laughter). The responses differ from those of the naïve pilot. We call the pilot naïve because we assume that he or she does not know that there are humans existing in those dots that he reckons move randomly. It is plausible that he might call them particles, which is no different from what we have called the particles that we appreciate through the use of powerful microscopes. But how can we know that their action is not random? What if it is all coordinated, or even purposive? The pilots now know that the dots they see below are purposive. I hope to guide you in these latter chapters to appreciate that the particles are at least purposive, in the sense that they tend to avoid annihilation. I shall then continue to explain, as I have so far done, why they need to be purposive in this sense.

Thus far, we have only tackled two of the four S's of the arrival of the fittest. The third is one that we have already encountered, but have not dug deeper into – self-organization. This property of an organism will bear relevance not only in an atom but also in a system such as earth. Corina Tarnita has shown how self-organization patterns can be qualitatively predicted in the fairy circles found in Sub-Saharan Africa.[21] The quantitatively predictive bit of the theory has to significantly stem from the last S – Size. While I came to only consider OS as worthy of qualitative prediction, size and to some extent, self-organization, also showed that it can have quantitative predictions too. Size is the one factor that completely distances OS from Natural Selection. For instance, while Natural Selection attributes reproduction to an organismal need to enhance its genetic prowess, or for a gene to multiply in the gene pool, OS attributes it to size. Size, once it has achieved criticality, what Per Bak calls 'self-organized criticality', it results in an avalanche. It produces, what is otherwise understood as reproduction. Reproduction, then, is an event that results from criticality. Coupled with self-organization, size gains traction in explaining the quark, the amoeba, the blue whale, the tribe, the city, the earth and even the observable universe. These are aspects that we will dive into in the next series of chapters. Buckle up!

Self-organization

It's like a jungle sometimes it makes me wonder how I keep from going under – ***The Message, Grandmaster Flash and the Furious Five***

World can't hold me, too much ambition, always know it be like this when I was in the kitchen ***– Next One, Jay Z***

The configuration of the hydrogen atom is such that is has an atomic mass of 1 – one proton. On its outer boundary is a cloud (space) but with only a single electron. The proton has a positive charge, while the electron has a negative charge. This dynamic relationship makes up the hydrogen atom. In this cloud or orbital, there appears to be some space for one more electron. Thus, the atom has a capacity for growth. Should it encounter another hydrogen particle with similar feature, they can merge and form the hydrogen gas molecule. One particle will bring its electron, the other will bring its electron, and they would dangle in this electron cloud space that can accommodate two electrons. The two atoms have capacity to self-organize past a certain energy threshold, which result in the formation of a new organism – the hydrogen molecule. The result is a structure whose order is generated by the interacting particles that constitute it. The properties of the molecule, the hydrogen gas molecule, are emergent from these interactions. For instance, there is the formation of the hydrogen bond which is non-existent in a singular hydrogen atom but exists in a hydrogen gas molecule. This much is evident in the particulate organism. What of a cellular organism?

The *E. coli* has the capacity to replicate, again after it meets a certain threshold that points towards the maturity of the bacterium. One can imagine a growing blob that at some point breaks into two blobs. From an initial entity follows a bifurcation into two entities. At its initial phases of *E. coli* replication, it has to go through clearly-outlined stages of cellular division. It has to duplicate its genetic material in the cell's nucleus, which separate by going to opposite ends of the cell. It then has to develop a constriction that cuts the cell at just the right angle so that there is an identical cell after the split just like the prior one. It generates another organism, after it has organized itself in different intricate steps of cell division known as binary fission. The resultant cells have features that are strikingly similar, resulting from the interactions of constituent particles forming the respective bacteria. Compared to the atom, the capacity for self-organization in the *E. coli* is astounding. Even more so is that it can do this about 70 times in a single day.

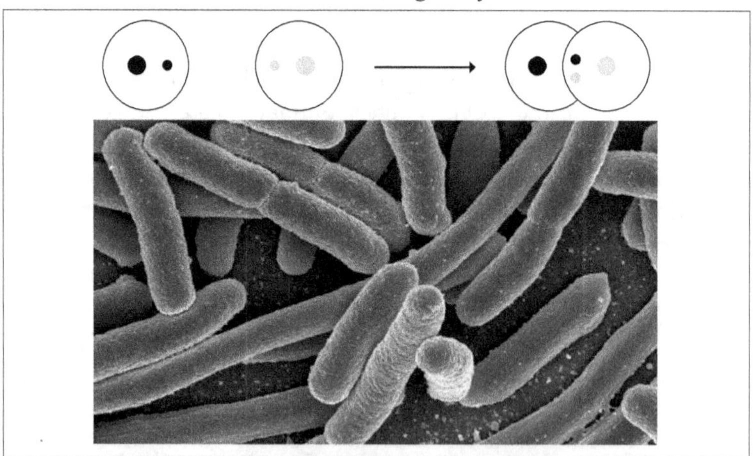

The different levels of self-organization. The upper image shows how two hydrogen particles, each with a single electron and proton, can merge to form a hydrogen gas. The lower image shows the rods of *E. coli*, capable of growth and multiplication.

Are there any other structures that can fall between one extreme of atoms and the other of a bacterium? There are. A good example is Bénard cells. These cells can be formed from experiment, when you set a petri dish with a thin film of oil and subject it to heat by setting an electric pan under it. The idea of using such a pan is to ensure uniform transfer of heat. In a short time past a certain heat threshold, at the surface, there develops a mosaic of cells with clear boundaries and holes at the centre. Contrary to the idea of heat being a dissipative process, a process that distorts, it appears to create a structure – a dissipative structure. This is possible through a localised type of convection that happens on each Bénard cell – the Rayleigh–Bénard convection.[1] It is an example of a structure generated from a dissipative process. The concept of dissipative structures is one that the Nobel laureate Ilya Prigogine studied extensively, using this simple experiment.[2] Such emergence of a complex structure, in the presence of heat, hints at self-organization being present in not just living organisms, but in particulate organisms. Even in the presence of a destructive property such as temperature, there appears to be some degree of organization.

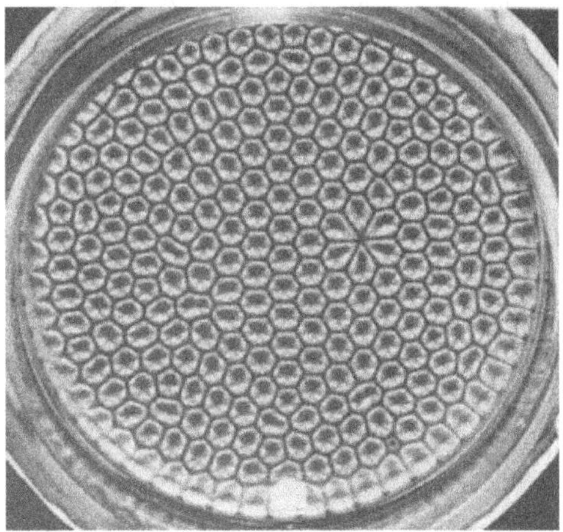

The hexagonal cells of Raleigh-Bernard cells. This level of self-organization lies between the two ends described in the previous image – between the hydrogen gas and the *E. coli.*

These three examples are different faucets of self-organization. Self-organization is the emergence of a globally coherent structure from locally interacting particles. It is a concept derived from the study of systems. Often, the systems in mind are those that are far from equilibrium, but not in chaos – systems at the edge of chaos. Just as in the examples given, there are various ways of looking at it, amounting to a multifaceted view at different scales. Furthermore, self-organization introduces the idea of the capacity for systems to self-organize that often works in terms of thresholds. Past a certain threshold, the hydrogen atom mergers with another to form a hydrogen gas molecule; the bacterium beings to replicate and; Bénard cells form. In the atom, its range appears very limited. In the dissipative structure of Bénard cells, they appear to form a concise structure despite the heat generated from it. In the *E. coli*, despite its complexity, it looks like it is the order of the day. Self-organization is clearly a feature of complexity.

There are key areas that we can focus on using these examples, the first of which is that the various systems are open. In classical physics, the development of the laws of thermodynamics insisted on closed systems. This step in the history of the subject proved instrumental in generating the currently understood thermodynamic principles. From this paradigm, there was a system, and its universe (sounds familiar?). Enter Ilya Prigogine who studied a system that was capable of self-organization, in spite of its subjection to heat. Heat is known for generating random movement, but the Rayleigh–Bénard convection and the resultant cells are not random. It is a consistent result tested by many scientists or interested curious minds. What is evident is that heat goes in from the bottom and significantly leaves at the top– it becomes dissipated. As such, it is an open system. In the process, it manages to maintain its organization. Its structure remains intact throughout this process. That is, the structure remains organizationally closed. Therefore, organisms are thermodynamically open but organizationally closed systems. Here is one way to look at it. Imagine a situation where **A** causes **B**, **B** then causes **C**, **C** causes **D**, and **D** causes **E**…all the way to **M**. **M** could then, for some strange reason, cause **D**. **D** would then cause **E**, all the way to **M** and then back to **D**. We manage to attain a closed loop – an organization that is closed. But since each of these entities require energy to maintain the loop, they have to be thermodynamically open. This is the only way that allows for them to be organizationally closed. The locally interacting components generate the global structure that is a closed loop. The threshold that has to be met is the causative link that has to stretch until M. Otherwise the closed loop might not have formed. This is self-organization. Notice how is operates in thresholds and has a spontaneity to it. It is not known why M reverts back to D, but somehow and spontaneously so, it does.

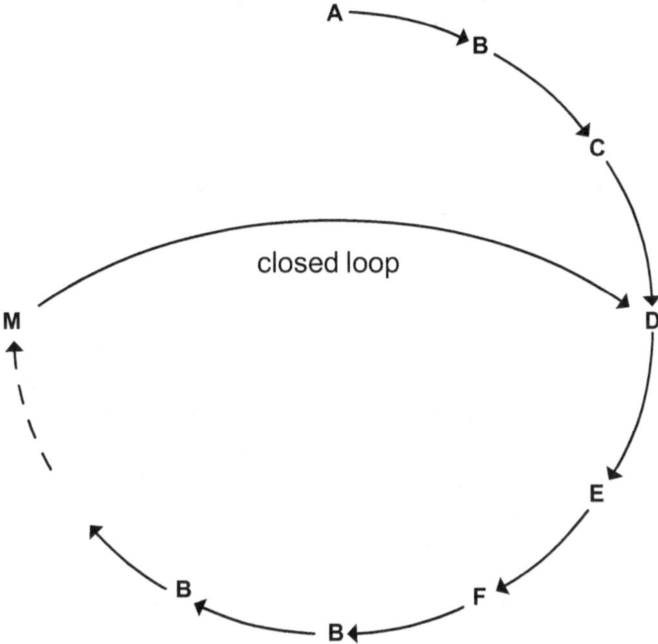

Formation of thermodynamically open and organizationally closed structures. Energy is needed for **A** to impact **B**, hence thermodynamically open. An organizationally closed system is formed when **M** ends up impacting **D**.

In the hydrogen atom, an electron from another hydrogen atom can penetrate the other's electron cloud or vice versa. The hydrogen atom is an open system. However, not all structures can easily penetrate the hydrogen electron cloud boundary. For instance, one of the most abundant elements, Helium, cannot easily merge with hydrogen. The hydrogen particle can, through the threshold effect, among other properties, weather a lot of external perturbations with very few being capable of penetrating its electron cloud boundary. Thus, for a while, it can remain organizationally closed. The *E. coli* needs to convert environmental substrates into energy and release the waste as heat before it can duplicate. It strives to maintain a closed organization by being a thermodynamically open system. All the three examples, the atom, the Bénard cells and *E. coli* are all thermodynamically open albeit organizationally closed structures. Yet again, at all scales, we have self-similarity, a fractal pattern of behaviour. Taking the map we developed, of an organism and its universe, the *only* way the organism can survive is if it is an open system. It would then seek those particles that it can merge with.

The openness then hints at the organism's *capacity* to self-organize. Self-organization would also mean that it cannot merge with any Tom, Dick and Harry. It manages to remain organizationally closed in this sense. Remaining organizationally closed also hints at an organism striving to maintain its structure in spite of its harsh universe. It can weather several perturbations by operating using a property known as power laws. Power laws imply that significantly large events are rare but small events are numerous. Translated to the life or existence of an organism, it means that it encounters very many small perturbations and very few significant ones. For instance, the rare and significant large perturbations happen when an organism finds a suitable merger or it becomes annihilated, with reference to the time of its existence in its universe.

The other key aspect is that self-organization demands that energy be transmissible throughout the system. There should be various avenues where the energy can traverse the system. In the atom, electrons are mobile in their electron clouds, or more specifically, in their various orbitals. As for the atom itself, even after merging with another atom, it is still in a state of continuous motion. This state is indicative of the transmissibility of energy throughout the atom. For the Bénard cells, there is the transmission of heat through the Rayleigh- Bénard convection. For the *E. coli*, there is the energy generated from substrate metabolism. The state of a system and its capacity to self-organize is a function of transmissibility of energy. Related to this is that the degree of self-organization is varied at different scales. The formation of the snow flake differs from the formation of methane. The parthenogenetic (self) reproduction in some insects, plants and certain reptiles differs from the sexual reproduction seen in mammals. Across different scales, there are different degrees of self-organization. These variable levels of self-organization dictate, significantly, the evolvable capability of the organism. For the hydrogen atom, there is an empty slot in the orbital (electron cloud). For the fruit fly, there are several options, from the egg state to the adult stage. It can therefore be taken as a rule that the self-organization capability of an organism determines to a great degree the *evolvability* of that organism.

It is important that there is this asymmetry in self-organization, as it is through asymmetry that we appreciate evolution. The asymmetry can become dynamically large as the organism increases in size. It could even allow an organism to make minor or at times major shifts in its internal constitution just so that it can continue existing. Different scales of the organism would then translate to different forms of diversity, variability, and strategies to avoid annihilation. From self-organization, evolution springs out. Ergo, even if there is a single particle, with unscaled variability, it is bound to be evolution. Coupled with variable forms of resistance to perturbations, self-organization would explain how antimicrobial resistance arises in spite of the numerous health advances humans achieve. It also helps us appreciate the technological leaps by humans and other animals since this in itself is a form of self-organization. The very idea reminds me of a very insightful manuscript I read by Ainissa Ramirez.[3] Ainissa Ramirez is a materials scientist and science communicator. She is passionate about teaching science so that people find it not only fun, but impactful to humanity. On the Edge website, she explains how the various materials we create shape us. To make this point, she gave a story of Ruth Belville, an expert trader in time. Before this niche was exploited by watch-making companies, Belville would make sure that she had her watch adjusted to match that of the Greenwich Meridian Time every morning of every working day. She would then move from office to office getting a fair trade for the truthful time in London. Companies were dependent on her and she was dependent on them. After companies sprung and the watch-making business boomed, humans ended up being significantly shaped by our clocks and watches. We then began insisting on punctuality, always asking what time it was in the day and even making tight, time-bound schedules. Watches and clocks, considered inanimate, shaped us.

What of animate organisms? When a mother is pregnant, for the first time, they might not know how to handle or even raise the baby. Once the baby is born, she picks up skills on how to latch it on the breast, how to respond to various stimuli the baby produces, a cry, a foul smell or even the grasping reflex it has in the early days out of the womb. The baby shapes the parent. The parent, likewise, shapes the baby. When the parent has some errands to run, she assigns somebody to temporarily take over her duties. The village or surrounding society thus shape the baby and the baby shapes the society. It even extends to other inanimate organisms. The oxygen determines the growth of the baby and the baby determines the cumulative degree of oxygen in the atmosphere in different locations. The hoe shapes the farmer over time much in the same way the farmer ends up giving the hoe a different shape over time. In this sense, isn't the inanimate any much an organism as any animate one? Don't they shape each other?

Here is another example of light, which can also be considered a particle. A lit candle transmits less of blue light, which signals the brain to switch off its systems and head on to sleep. The invention of the incandescent light, however, emits a lot of blue light, which gives the body a signal that it is still daytime. Such signals have an effect on hormones produced, such as growth hormones, the result of which blue light exposure enhances growth. Light, considered inanimate, shapes the animate human being. The inanimate organism ended up shaping the animate. If the light we currently use is different from the candle-light that was the norm many years back, it only speaks of the lifetime of the 'inanimate' organism relative to the 'animate' organism. All these organisms shape each other. While this dichotomy of animate and inanimate is useful in distinguishing some similarities, it could also blind. Each of these organisms has different self-organization capabilities, amounting to different asymmetries, allowing different paths of evolution to take place.

The relationship between the animate and inanimate is also the story of self-organization. Considering the map of an organism and its universe, the organism that escapes annihilation is one that can self-organize to harness what its universe contains. The human would then increase in size when exposed to a lot of blue light. It would then have to get more food for itself – an unforeseen outcome in the organism's universe. It is the unseen problem, P_2, to the solution for P_1.

As a result, an organism ends up shaping itself in its own universe, depending on its capacity to self-organize. For instance, the human being is perceived to have a high degree of self-organization compared to a spanner. Thus, the features of what biology would call an organism are attributed to the high level of self-organization capabilities compared to the one that has little. However, a spanner can weather a wide range of perturbations, arguably much more than the human being. In fact, it is hardly ever appreciated that such a capacity exists for an object such as the spanner. Upon high resolution, this capability is present. The spanner has mobile atoms. It also has electrons dancing in their various orbitals. Energy can move across the object the same way energy traverses a leaf under sunlight. It is able to remain organizationally closed much easier than a human being. When you stroke a magnet along a steel spanner, it remains a spanner, but shaped by another organism, the magnet. Superficially, it looks like a spanner, but by its internal constitution, its particles are aligned such that it attains new properties – magnetic properties. It is able to self-organize and acquire a property it did not initially have in an 'unmagnetic' state. Self-organization is a feature evident across all particles, and as such, all organisms. It is the property that equips them with a slight edge of avoiding annihilation in their respective universes. Specifically, each organism has that power law property working through threshold effects to avoid variable universal perturbations. Hence, the simplest organism is a particle and its tendency to avoid annihilation in its own universe.

Self-Organization

The idea of self-organization introduces the concept of self. Not self as psychology has always variably described it, but self in recognizing a degree of coherence among the locally interacting particles of a system. For there to be self-organization, there has to be an evolving self, in the general sense of the term, to organize. It is another marker of the presence of an emergent organism. It takes us back to the rather vague but convenient term – merge. As highlighted previously, merging differs are different scales of an organism. At the sub-atomic level, it comprises strong nuclear forces. At molecular and atomic levels, there are bonds constituting steric and electrostatic complementarity. At cellular levels, there are junctions and signal pathways. At organ and organ system levels, here are duct systems. At the human-human level, whale-whale level, bee-bee level, there is language or communication. All these mergers result in emergent organisms.

Merging demands and results from a significant degree of coherence. For instance, when atomic particles emit light in a system open to continuous direct current, the atoms behave like a lamp, with variable degrees of coherence. Past a certain threshold, the light changes dramatically and qualitatively, creating a singular wavelength. This is the mechanism of the formation of laser light.[4] The emergence is evident, but only where there is a substantive degree of coherence. When compressing gas in a chamber, the gases are variably incoherent, with their solitary movements. However, as the compression continues, it passes a critical point where there is a qualitative and dramatic change – the gas liquefies, attaining emergent and new properties. From the random movements of the gas particles emerges a coherence of the liquid particles. It is unpredicted, unforeseen by the organisms but it happens nonetheless. This is the power of self-organization. When a man meets a woman, they individually use their ability to understand each other's sound vibrations and signals to self-organize, and create a merger – a couple. Later on, they can merge and form another being, their child, who at first, does not know how to coordinate his or her vocal cords to match those of their parents. The parents also do not know how to make meaning from the invariable rants of the baby. With time, each enhances each other to the point of coherent communication. The merger has resulted in another organism. It might appear strange that a family is considered an organism. It only speaks of the variant means of merging and the disparate capacities for self-organization. How?

If there is a merger, there is the formation of a new organism. The type of merger formed depends on the two or more original organisms, whether particulate or animate. How does one then know that they have formed a new organism? Coherence of the particles might not appeal to many as indicative of the emergence of a new organism, even though it is evident. Coherence is what some might call order that stems from chaos.[5] Nevertheless, I have found one single means of detecting the presence of an organism, which takes into consideration all types of mergers that can be formed. This litmus test is: a form of credible, imminent danger or threat. If one seeks to identify the presence of an organism, one needs to expose the organism to a credible form of threat. Not just any type of threat, but a credible type of threat. A credible threat is potent enough to strike the organism as to either distort or assert its organization. The threat constitutes a credible perturbation to which a system would respond by resistance. This is what I mean by *self*-organization. *Self*-organization is how the organism behaves or acts in order to resist annihilation.

Let us explore this from one scale to another. At the sub-atomic level, we have the organism that is formed by quarks. The most stable one being the 'tripod' quark, with three constituent particles. The credible form of threat we can subject this organism to is to separate one particle from the other two. This is met with a lot of resistance. The further the distance of separation, the greater the resistance. This is a quark feature known as asymptotic property. A scale higher, at the atomic level, there is a nucleus and a cloud of electrons. Attempting to separate the atom from the electron requires energy. This type of merger is the electrostatic force between the protons and the electrons. Slightly above the atomic level, that is, the molecular and ionic level, there are various forms of mergers – hydrogen bonds, ionic bonds, covalent bonds and even Van der Waal's forces. A scale higher and we can consider sand. Sand particles may appear to have no connection when in essence, they do – weak gravitational force. Two bodies are always attracting each other with a force that is inversely proportional to the distance separating them. Separating the sand particles requires energy. But it happens so easily that one might think that such an attractive force is absent or non-existent. Gravitational force is weak if the mass of an object is small. Nevertheless, there is some resistance that one has to surpass in order to overcome the attractive force of gravity. What of the bacterium? It is the hallmark of the rapid rise in antimicrobial resistance. It would try to resist the threat that antibiotics present.[6] The same goes for antiviral and antiprotozoal medication. What of a family? Typically, try attacking any of the family members and you will be met with the kind of force that solidifies family dynasties. Try taking a baby from a mother and witness maternal fury. My elder brother has a scar on his forehead when he tried taking a chick from the mother hen (or so the story goes), and the mother did not care for the large size of my brother back then. It attacked. Similarly, try attack a particular race and you will spark a revolution. Try to attack a particular gender, and what you have is the formation of a movement. The single most accurate means of detecting the presence of an organism following a merger of whichever kind, is subjecting the organism to credible threat to their existence. Why? Because an organism is particle or particulate organization that seeks to avoid annihilation. In response to this threat level, it will *self*-organize.

Self-organization, therefore, exemplifies various forms of mergers. A unique and beautiful example is seen the Redwood tree. The majestic Redwood can live up to an astounding 2000 years! But the bulk of its 'living' self is a thin layer of cells just beneath the bark of the tree.[7] Once they die, they contribute to the bulk of 'dead' material that makes up the main stem of the tree. The organism in this sense is the tree, comprising mostly dead material and very little of the active cells. Looking at only the living cells narrows one's scope in appreciating the contribution of the 'dead' cells to the constitution of the organism. Even the human body has a series of dead cells. The skin, the largest organ of the body, has an outer layer made up of dead cells. This means that a significant portion of the human body comprises dead material. They, however, serve a role in the very organism that we call the human being. It should strike one as odd how the constitution of dead particles and dead materials should, past a certain threshold, become all of a sudden, alive. Past a certain threshold of light-emitting atoms, we have the laser light. The laser light is however, not considered to be alive. Granted, it is the recognition of a unique organization of particles that makes us consider certain systems as alive and others rather not. Even though such an organization might be called upon to distinguish living creatures from particulate ones, it is circular. That is, the mere existence of a particle shows that it possesses a unique form of organization that manifests after exceeding a certain threshold which allowed the particle to exist as it does. Why not simply consider all particulate material organisms? It dissolves so many arguments while still saving the various disciplines of 'inanimate' physical sciences and 'animate' life sciences. Self-organization thus springs up novelty beyond the simple structure that we see. From an existing particle, it allows us to appreciate *its* coherence. A particle would quickly disintegrate if it were not coherent in its structure. Coherence implies that a certain threshold has been met. It further implies the existence of mergers which are better appreciated through the concept of hierarchies. It is hierarchies that elaborate why the Rosewood tree would still hang on to the bulk of 'dead' material, much the same way as we stick to our 'dead' skins for survival. So how do hierarchies, as understood in systems dynamics, work?

The transverse section of the Rosewood tree. The thin outer bark comprises the living cells that add to the girth of the tree. The remaining chunk largely consists dead cells.

Hierarchies and Resilience

In systems, the concept of a hierarchy is as simple as the term itself – it defines the different levels of operation of an organism. As a medical student and human anatomist, I have a fair understanding of how the human body functions. I shall therefore use it as an example. In ensuring we get to understand how hierarchies work and in the interest of preserving what Biology has taught us, we can consider the cell as the lowest level of operation of a living entity. It is regarded as the basic structural and functional unit of life. The cell of a human being has its own organelles which it controls and orders following various signals. It can give instructions to produce a protein such as a hormone or a chemical such as gastric juice. A group of cells, whose function is synchronised and synergistic, forms a tissue (coherence). The constellation of tissues together forms an organ (coherence). Together with other organs, various organs form an organ system (again, substantive coherence). The communication between different organ systems forms the individual organism (don't they all cohere?). The cell is at a different hierarchical level from the tissue, and the tissue from the organ, the organ from the organ system and the organ system from the individual organism. These are all different hierarchical levels signifying different levels of operation. However, there is more to hierarchies than this rather straight-forward example.

The stomach cell of the gastrointestinal system is very ifferrent from the nervous system's brain cell, visually and even functionally. You would never imagine that they are related were it not for the cells of another system – the circulatory system. You could even say that the circulatory system introduces coherence between these two disparate systems – a form of coordination. These systems are bound together in a rather squished space by the appendicular system, comprising the skin and its structures. These systems do communicate through different pathways, but not always. The level of communication within a cell and its organelles is much more intimate than that between a cell and another. That between the cells of the liver is much more intimate than that between those of the liver and the stomach. That between the cells of the gastrointestinal system is much more intimate than between the gastrointestinal system and the respiratory system. In short, what hierarchies do is minimise the need to have one system keep constant track of another system distant from it. It is one of the reasons complex systems work well. I do not have to know the weight, height, hobbies, fears, or even the radio channel that my neighbour is listening to. I only need that which is necessary at different times. Often, that will largely be his or her name every time we meet.

Hierarchies reduce the amount of information that a system has to keep track of.[8] It is why a family works. The level of communication between the parents is different from that of the parents and the children. Even among the children themselves, it differs based on various aspects, such as age groups or even sex. In this hierarchical system, there is an emergent organism that we dutifully call a family. It comprises sub-systems – the members of the family. In each member, there comprise other sub-systems, which are the organ systems of the members of the families, all the way down to the sub-atomic particles. In all these multiple hierarchical levels, I do not have to keep track of how my atoms are vibrating. For a long time, humans have not known that they have had genes, but they have still been functioning. I am fortunate enough not to keep constant track of how my kidneys a functioning. In short, what hierarchies do is reveal the level of *autonomy* of every system at every scale. The autonomy espouses the very concept that considers particles and in general, systems and subsystems as organisms.

Various studies of complex systems have focused on concepts such as the degrees of freedom in systems and enslavement, but the other flip side of it is the presence of autonomy. How else can enslavement exist without autonomy? It is the autonomy that allows us to consider the existence of an organism in the web of interconnected hierarchies. The proton has some degree of autonomy as much as the electron. A merger between the two constitutes the bond that they have made – an electrostatic bond between two autonomous organisms. The result is another emergent organism with its own degree of autonomy (and disparate degrees of freedom at this hierarchical level). The autonomy of the emergent organism explains why the electron resists separation from the other constituents of the atom since the sure way to test the existence of an organism is by subjecting it to some form of credible threat.

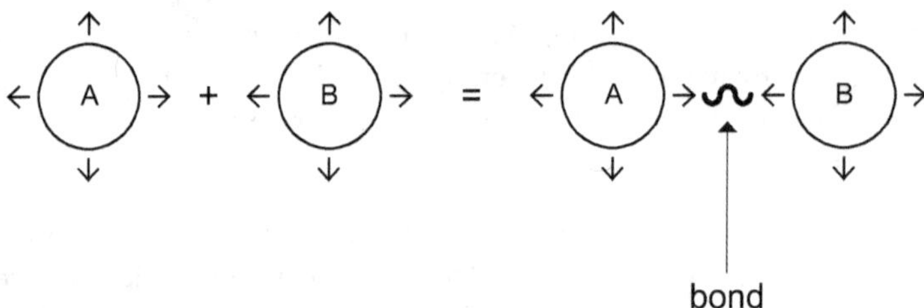

Degrees of freedom. By themselves, individual particles display more degrees of freedom when separately added. However, when these particles merge, as in the case of A and B merging, these degrees of freedom reduce. This image illustrates the process of enslavement among otherwise fairly autonomous organisms.

In addition, the ability to have these hierarchies streamline their goals and the ability of organisms to self-organize contributes to not just stability but resilience. Resilience is another concept in systems dynamics that defines the playing ground of a particular organism.[8] It highlights the ability to integrate new particulate organisms, and readjust to the flux in the organism's universe. The flux is what might be technically regarded as 'noise' or the perturbations that organisms are ever subjected to. As a result, resilient systems have various feedback mechanisms whose roles protect the organism. Feedback mechanisms stem from links between and among every autonomous organism at each scale and the hierarchical system that constitutes organization in an organism. The ability of an organism to adjust oneself through the feedback mechanisms and existing hierarchical systems so as to escape annihilation then comprises self-organization. It then becomes clear how self-organization, hierarchies and resilience are tightly bound. I could not have expounded more about self-organization before touching on hierarchies and resilience. In the interest of clarity, here is an example that illustrates their interconnectedness.

The human body temperature is unique in the animal kingdom – it is fairly invariant, ranging from 35^0-37^0celcius. At times, these extremes of 35^0 and 37^0 can be considered as cause for alarm by physicians in different conditions. So, we can safely state that the average temperature is 36^0celcius. As medical students, we have always been taught about a particular set point that determines the body's temperature. In retrospect, I warmed up to this explanation in spite of the fact that the human temperature is a range and not a point. Further, when a patient has fever, the explanation has always been that there are some chemical messengers (prostaglandins) that are produced and reset the set point somewhere high enough to increase the body temperature just so that the individual can survive the infection while preserving its own life. While this hypothesis has set camp in medical school, I later came to question its validity. True, prostaglandins are involved in the fever pathway. However, the idea of a set point is limiting – it does not take into consideration that the human temperature is a range and not a point. Even during moments of fever or hyperthemia, the temperature is still variable (elegant experiments have, however, shown that there are cells in the anterior hypothalamus, the region just above and behind the eyeballs, that do play a role in temperature regulation through the process of negative feedback).

Bearing in mind the degree of autonomy at each hierarchy, I subscribe to a different school of thought that I believe is much more interesting. There are independent mechanisms that contribute to the overall stability of the temperature seen in the human body. It is courtesy of Lovelock that I stumbled upon this idea; how there are five central mechanisms that regulate body temperature. These are: core-modulated shivering, skin-modulated shivering, cutaneous vasodilation, sweating, and biochemical heat production.[7] Each of these responses from different systems are activated at different moments, signalling their autonomy and collectively, the body's resilience to variable environmental changes in temperature. The resilience further is captured by the diversity of the sub-systems at different hierarchical levels of operation working through feedback systems that might be linked to the anterior hypothalamus (which is the highlight of the set-point hypothesis). The net result is a regulated body temperature. The narrow albeit stable temperature range speaks of the coherence of the human being much like the coherent wavelength of laser beams. The precision falls within the range of 35^0 and 37^0celcius, the normal body temperature.

In support of this argument is the ability of the more distant part of the body, such as hands and feet, to have extreme temperatures which can fall outside the normal narrow temperature range. This evidence further brings to question the set-point argument of temperature regulation. I would wager that this is the same mechanism that holds for birds, who are also homeotherms (creatures with fairly invariant body temperature). For the poikilotherms (creatures whose body temperature varies with the environmental temperature), they have to modulate themselves by either seeking warm or cold surroundings. Each of these actions displays degrees of autonomy, at different hierarchies, constituting self-organization and demonstrating the resilient capability of an organism. The interconnectedness of self-organization, hierarchies and resilience is inescapable.

Furthermore, self-organization can happen in any of the hierarchical levels, as long as it stands in line with the goal of the system – that is, the goal of the organism. From OS, the goal of the organism is to avoid annihilation. Therefore, at the organismal level, it will carry a gun, learn karate, walk in groups, spit venom, act dead, speed off, hide in its shell or even attack another organism to avoid annihilation. At the organ system level, it can encourage hypertrophy of muscles or even create a hormonal cycle that rewards the organism for engaging in social interaction – this is the hormone oxytocin, or prepare the body for an energy-demanding process by releasing cortisol. At the cellular and sub-cellular levels, there will be a storage of energy rich molecules and their subsequent release when the need arises. Self-organization at each of these hierarchies makes sense when aligned to the goal of the organism, which is, to avoid annihilation.

Of special note is that self-organization does not at times display immediate outcomes. A business does not immediately announce to the public that it has changed its computers, or that they are now using two UPS systems rather than three. The bacterium does not announce to the other bacteria that it has altered its make-up due to changing times. It, can, however, spread this innovation through horizontal gene transfer, only after it has fairly stabilised it in its genome.[9] Before then, the change is only known to the bacterium, to the organism. It is in fact highly plausible that the bacteria might not be aware of the genetic shifts, since hierarchies ensure that there is little need for tracking full information at the sub-system level. It should therefore not be surprising that there are different genotypes in different organs. The different post mortems that reported different karyotypes in different organs is just self-organization at work, in one hierarchical level. Since self-organization happens in different hierarchies, it should also happen at the genomic level. As a result, self-organization introduces particulate asymmetries. Asymmetries account, in part, for the diversity of the organism. The asymmetry further captures the unpredictability of the organism's fate, and hence accounts for the very existence of evolution. Thus, evolution will happen at the different hierarchy levels with variant self-organizing capabilities. As I will show in a little while, self-organization is also evident in language, but to my knowledge, this property is hardly ever been recognised as such.

Hierarchical Autonomy

I have defined evolution as the ability to avoid annihilation – genetic mutation, epigenetic inheritance, cellular hypertrophy, autophagy (self-consumption), systemic blood-shunting and organ system coordination, fighting or fleeing, staying in residential neighbourhoods, developing a standard means of communication such as English, French and Swahili – all these are different forms of self-organization at different organismal levels. Since hierarchies depict autonomous systems, evolution happens in part when organisms self-organize. Genes are therefore not dictatorial masters of the shell (organism), but autonomous organisms in the same way the quark, the electron, the proton, the cell, the tissue, the organ, the organ system, the individual, the family, the society, the community, the city, the company, the country, the race, the gender, the earth, the star, the solar system, and even the galaxy is. And why are all these entities organisms? They are all organisms because if you introduce a credible threat, it will be met with resistance!

Let's take it up a notch and dive into some specifics. It has always been known that the genome one inherits from their parents is supposed to identify a single organism from another. All of about 37 trillion cells of a human being are supposed to have this specific genome. Well, this archetype was empirically debunked by several studies. One of them showed that the brain cells of 37 out of 59 females had the Y-chromosome-specific DYS14 gene.[10] The Y-chromosome is often used as the genetic marker for maleness, but a gene formed by this chromosome was found in roughly 63% of the females in this study. It begs the question – where did a Y-chromosome-specific gene emanate from in a genetically female individual? The same DNA structural variants were seen in other organs such as the skin, blood and heart. Yet another study by researchers at Yale, found that a high fraction of children with congenital heart diseases carried mutations not present in either of their parents.[10] This could mean that DNA that we acquire after fertilization is not structurally persistent throughout life; it changes! Such evidence cracks the monumental edifice of the idea of a causal link between the inherited parental genotype and its phenotypic expression. The body is actually a complex adaptive system with self-organization capabilities that extend beyond the idea of evolution using a common genetic toolkit. Genetic mutation is just but a small facet of self-organization! Each hierarchy level works dependently albeit as autonomous entities, which allows for the manifestation of these varied genetic pictures.

The autonomy of different organisms would then account for the property of self-regulation attributed to organisms, biological or otherwise. The diversity makes up the reinforcing and balancing feedbacks seen in different sets of creatures when they are viewed as systems. The Gaia Hypothesis of Lovelock showcases this exceptionally. Lovelock beautifully explains how there are various cycles in the earth, such as the hydrogen and carbon cycles. Each has a degree of autonomy to it. Nonetheless, the various interactions constitute the balancing feedbacks that maintain the fairly constant percentages of the composition of air. What was usually stressed when I was in primary school was that Oxygen is approximately 21%, Nitrogen 78%, Inert gases at less than 0.97% and Carbon dioxide at 0.03%. Unfortunately, the same class environment hardly shaped me into asking why the disparate compositions have such a constant percentage. I got the answer from the Daisyworld Model, which, again, hinted at the degrees of autonomy of systems.

Lovelock is responsible for developing the Daisyworld Model that brilliantly showed how the autonomous processes of two types of daisies, black and white, can yield a fairly stable temperature of the Earth. In the model, there are no clouds, so that there are no other buffers to solar output. The black daisies absorb heat and increase the overall heat of the Earth in response to the sun's radiation. The white daisies do the opposite, they reflect the solar output. The progressive evolution of such a world is the arrival of a fairly constant global temperature. Variety yielded stability. Since then, various models have been developed with more species besides the black and white daisies, with the result of more stable parameters, such as a minimally fluctuating global temperature. This is an emergent feature, generated without any foresight nor any intentions; just autonomous systems about their drive to avoid annihilation. The more the autonomous systems were added, the more Daisyworld became resilient to perturbations. With the various hierarchies of the bird or human body, it then does not come as a surprise that mammals and birds have fairly constant temperatures. It should also not surprise us why reptiles, fish and amphibians do not have this property generated autonomously – there is no foresight. However, once we incorporate their behavioural tendencies to regulate their body temperature, these 'cold-blooded' animals also have a fairly stable temperature. The extremes of temperature are a form of credible threat to the frog, fish or snake. Behavioural tendencies are feats of self-organization at the hierarchical level of the individual. They generate a fairly stable pattern of behaviour and hence a fairly stable temperature.

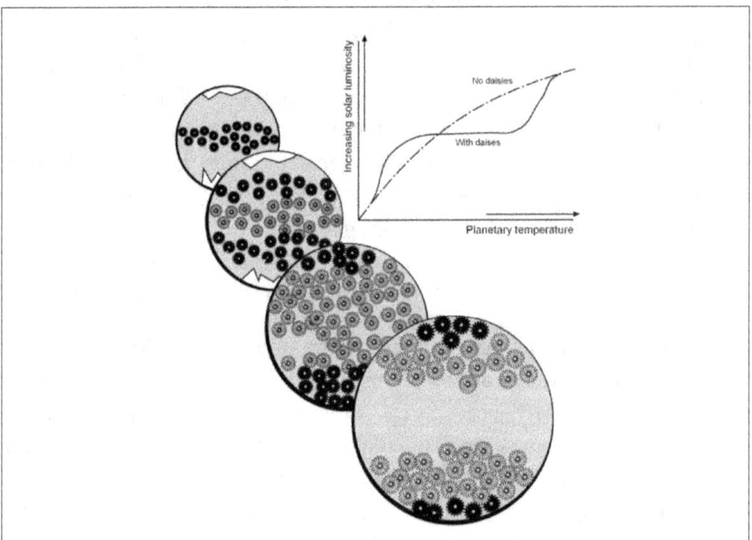

The Daisyworld model. The daisies have an effect in controlling the overall temperatures of Daisyworld through feedback mechanisms. This model was first formulated by James Lovelock to demonstrate the self-regulative processes of the Earth.

Maintenance of a fairly stable temperature strikes at the heart of self-regulation. Do atoms possess such a trait? I believe they do. The questions that led Lovelock to discover the Gaia Hypothesis are similar to the questions I asked myself about an atom. Why is it that the atomic structure is fairly constant despite universal perturbations? These, I concluded, were possibly evolutionarily stable properties that are indicative of a self-regulatory organism. The atom contains an electron, with properties different from those of its nucleus, which comprises protons and neutrons. These entities maintain the universally acknowledged parameters such as melting/freezing and boiling points. They possess these fairly constant parameters that emerge from the independent properties of the electron and the nucleus that gives the atom stable-enough properties for chemists and particle physicists to study. Such a balancing process is also seen in the stars, which have a dynamic stability between the fusion generated by strong attractive gravitational forces and the repellent energy generated by the nuclear fusion. Self-regulation is the ability to maintain a dynamic equilibrium in spite of the autonomy of the different constituents and their properties. It is similar to the self-regulating property of warm-blooded organisms such as birds and mammals. Why, then, not consider atoms as organisms? Is it because they do not reproduce? I shall settle that concern in chapter seven.

Up to this point, I have subtly if not conspicuously used concepts from systems dynamics. I am therefore compelled to highlight one thing that drives systems – their goal. I never would have known the power of a system's goal were it not for Donella 'Dana' Meadows. Dana fit all my mental models into one lattice of invaluable utility after I read her book, *Thinking in Systems*. She introduced me to the idea of leverage points and how they can not only be utilised, but how they are often counterintuitive. The theory I propound looks a little counterintuitive, but its ability to explain so many concepts was enough for me to pursue writing this book. But for now, let us head back to systems' goals.

The goal is powerful enough to shape how an organism behaves – it dictates its self-organization. In Dana's ranking of leverage points, self-organization falls immediately below the goal of a system. In relation to

OS, the goal of every single organism becomes pretty simple – to avoid annihilation. Since self-organization is not as powerful as the goal of the system, it would then mean that an organism would self-organize to avoid annihilation. I do not think that there is any other trait seen in living creatures that explains this concept more mysteriously than the idea of language.

Grasping some central tenets about language opens a portal to understanding a lot about what I have already described. It will support the map that I have hopefully by now, lucidly drawn – the organism and its universe. It would then highlight the powerful capability of the human being to self-organize. This will be clear once we acknowledge that self-organization yields a stable pattern – it has the potential to yield order from the chaos that autonomous organisms bring. For instance, it results in a stable global temperature in spite of the extremely cold poles and the hot tropics. It allows for the atom with the negatively charged electron and the positively charged nucleus to result in a stable atom. What, then of language? Autonomy begets diversity and diversity eventually results in stability. The diversity of the DNA gives its fairly stable state. The same diversity that yields stability is also seen in language. I later came to discover a remarkable feature about language that highlights why the organism is the fundamental unit of diversity. Hopefully, by the time you will be done reading, you will understand why Judith Rich Harris insisted that there are simply *No Two Alike*.[11]

Language

When reading *The Language Instinct* by Steven Pinker, I came across a finding that puzzled me. Pinker quotes Noam Chomsky when he says that every sentence ever constructed by humans is unique and has never been constructed by any other being – it is unique in the universe. It got me wondering what that meant. I knew so many sentences to be repetitive – just go to an elementary school! What was the basis of such a statement? I could only surmise that the string of word combinations has the possibility of making each sentence unique, so that uniqueness could be a function of probability. Nevertheless, should this be true, it has the potential for supporting our robust map of the organism and its universe. My guess may be biased, in line with my theory in asserting that Chomsky might be right. I am; however, sure the acclaimed linguist must have done his analysis before uttering such a statement. On the other hand, I tend to think that there are other ways we can deconstruct his statement and arrive at the same conclusion as his. We can do this by, firstly, looking at word organization.

If we were able to reorganize the 26 letters of the English alphabet, even if we had started the alphabet with B or Z instead of A, it would not have made much of a difference since words contain letters jumbled up in no clear rule for their designed order. However, word organization does make a difference. I can say **'Will you marry me?'** which to a literate person bears meaning. If I re-order the words, **'You marry me will,'** it sounds odd. Even **'Marry me, you will,'** feels like something only Yoda could say. Worse still, **'Me you marry will'**. It is not just the mere combination of words but the ordering that makes sense. In this example, I have given four possible combinations and out of them, only one appears to make sense (unless you have watched Star Wars). It is an example of how unique sentence construction is, from a sea of possible word combinations. These combinations increase exponentially with sentence length. From this sea of possible word combinations, very few strings of words make sense. In a sentence, words have a modular feature, just like the DNA, which accounts for another form of diversity. This is the first example of how unique sentences are or can be constructed. That is, there emerges coherence from a string of diverse words and their combinations.

Using the Popperian Schema, we see how the organism makes sense out of this tumultuous sea or words. It must first acknowledge the problem of a plethora of words that together just do not make any sense. This forms P_1. It then formulates a hypothesis of how they can generate meaning, being the **TT**, the tentative theory. It can only test if this theory has meaning from the response it gets from those superior to it, as a child learns from its seniors or from its peer, much like how Creole begins to form. It then does its correction, through error elimination, **EE**. In the process, it generates another problem of how to make meaning from the same sentence with the knowledge it has thus far acquired, P_2 or even, to repeat the same string of words once they have received feedback that the initial string of words makes (or doesn't make) sense.

The schema bears two important points. The first is that the organism generates its own hypotheses and tests them. In this process, it generates knowledge through error elimination and feedback. What it does for other materials such as food substrates, which are key to its survival, it does for language as well. Language is as crucial to its survival as material substrates. Here, I use language in the general sense of the term. Secondly, the unique sentences that a baby make are stronger evidence of the uniqueness of sentence construction than in adults. Children make a lot of 'grammatical' errors when learning to speak. So even their initial cluster of sentences in its early years will be unique and tied to the child itself – a unique production by a single child. This is the second evidence piece of evidence that can support the possible unique sentences constructed by every single individual. It also follows that language is not innate, but constructed. There might be some anatomical features that grant human beings with the ease of formulating different languages. However, that there are different languages is also supportive evidence of language construction and thus, of the process that qualifies the diversity albeit stability of different types of languages.

The third is the one that I found intriguing; accent. Accents are unique to one's place of origin. I was shocked to hear that the British have various accents, the Liverpool one differing from the one in London. Even in Kenya, you can tell a Kisii from a Luhya, just from the accent. Written words do not have accents, but once they are spoken the accents become evident. It got me wondering how the uniqueness of sentence construction is not just modular – it is not just a matter of how words are strung together, but how those words *sound* when somebody utters them. They are very different. Accents then led me to the various tonal variations. Tonal variation is a function of the voice box capacity, your nasal cavity whether clear or congested, age, position, audience, mood, affect and so much more. We can say the same joke, word for word, but the delivery is important. Hip-hop has majored on the delivery of punchlines, enough to make one a billionaire, as in the case of Jay Z, Sean Carter. The sound of the words and how they are voiced by an individual adds uniqueness to every sentence constructed.

Sentence construction has the unilateral tendency of being analysed as ordered words on a piece of paper or as a blank sheet in a smart device. But the type of voice also plays a role in sentence construction. I can say 'I'm the man with the bass' at the same time as my small sister. In her not-so-bass-like voice, the sentence has the same meaning, but is uniquely different with reference to other factors other than the words used. Some of these factors include the speaker's voice, with emphasis on particular tones, pitch, shape of the mouth, or mood. In fact, where you position yourself in a room when speaking can have a different resonance from another position in the same room. Indeed, sentence construction is unique in more ways than just word combinations.

The organism, in its universe, constructs sentences and tests them to survive in its universe. In my universe, I construct my own sentences and hope that there is meaning in them. The sentences are bound to be very unique in various ways. Taking into consideration the arguments I have raised, it appears that every human being displays an ability to organize words into a string that, more often than not, make sense. In spite the various 'errors' in language use, these 'perturbations' hardly affect the meaning of a sentence. Amidst it all, significant coherence exists. Coherence depicts stability. A stable pattern is the hallmark of self-organization. To sum it all, the uniqueness of every string of sentences further supports our initial map – an organism and its universe.

Judith Rich Harris put up a compelling argument of how language can be used to show that parents contribute less to the accent that children end up having. Say my family got to travel to England when I was a little child, around 4 years old. My parents, on the other hand, had lived in Kenya for the last 30 years before considering a change of scenery. We stay in England for 15 years before I even consider going back to Kenya. The accent that I am bound to acquire is the English type and not the

Kenyan type. The accent gets shaped by the friends that one interacts with more than with the parents. As an extension of her initial book, *The Nurture Assumption,* Harris shows the behaviour of a child is significantly shaped by their peers rather than their parents.[12] This was another curveball thrown at psychologists, who have built their foundations on the parental influence of a child's behaviour. Harris showed that this can be extended into Language, which is relevant in either excluding or including one in a particular group. In Kenya these two phrases 'pele-ka na rieng" and 'let us fika' distinguishes two separate groups. Choice of words depicts behaviour – language use is an extension of behaviour. If you behave in a certain way, you are included in a group. If not, you are excluded. The English behaviour and by extension, language, is very different from the American, hence the constant tussle between the two versions of English. And accent. And tone. And position. And intonation. If you have ever wondered why artificial intelligence (AI) has found it hard to capture the learning phase of the developing child? Behaviour is one possible answer. And behaviour defines language.

Long before language could even be thought of as an expression of behaviour, it was widely accepted that it comprised the transmission of information from one person to another. The then prevalent understanding followed from the idea that language bears information that is independent of any interpreter. This paradigm has been the basis for the development of computers and digital machinery, reliant on data as discrete as 0s and 1s.[13] It is this paradigm that fast-tracked the idea of creating an artificial intelligence (AI), as well as stalled the process of generating an artificial form of intelligence that mimics the growing child. Steven Pinker explains how it is these simple developing stages that prove the hardest for AI developers.

Transcending the paradigm of data as discrete and independent into contextual and behavioural might just be the elusive and much-needed piece to this puzzle. The fact that different people can get different meanings from a single sentence or that a word can be mispronounced or uttered in a different accent shows that language is contextual. The delivery of the sentence is important in defining the meaning of the sentence. This is in line with the idea of an organism and its universe, as the organism tries to make meaning of its universe from the numerous models that it makes. It tries to derive meaning out of the unique set of contexts it is always in. Using the guide of the Popperian schema, it has to first develop a tentative theory (TT) to solve this problem (P_1) it identifies, then eliminate errors (EE) before it advances to the next problem (P_2). In the case of language, the next problem might be the right pronunciation or intonation. (It might appear that there should be a theory present for an organism to identify a problem. In fact, there is. Remember, it must first have or create a tentative theory that does not match its reality, for it to recognise that there is a problem, P_1. The organism then develops another which if successful, is able to derive meaning from the data).

Communication as Coordination

There are different translations for the prayer 'Our Father'. Despite this, the prayer bears the same meaning. If one does not coordinate his or her tentative theory to match the problem that is the different translation of the prayer, one will not be able to derive any meaning from it. In short, language is contextual. An organism has to coordinate its tentative theories with its reality. When there is a substantive degree of coordination, the data then bears meaning. However, by coordination, I do not mean simple imitation, even though imitation covers some level of coordination. What I mean by coordination is the optimal or satisfactory execution of a strategy based on your immediate surroundings. For instance, when a snake meets a potential threat, it raises its head, ready to attack. The stance it takes is aimed at signalling to the potential threat that it will not go down without a fight. The threatened might also decide to attack or run away. If it runs away, the snake would then lower its head. If it doesn't, it will also take an attacking stance. The *behaviours* of the snake and the threatened signify communication. There had to be some level of coordination between the two for their respective models of their on-going reality to pick what the other was signalling. It happened through shifting of behaviour. Behaviour signalled communication. This changes the whole idea of language from discrete forms of data into contextual data and hence from discrete information, to contextual information.

Behaviour as a form of communication. In this image, the behaviour of the man might not be registered by the snake. In that regard, there may not be any form of communication between the two entities.

There has to be some level of coordination for there to be some level of communication. Different particles in a room behave in a manner as to signal the temperature of that particular room. The human being had the initial problem of calibrating a device (thermometer) to derive meaning from moving particles. Humans then created a tentative theory and actualised it by making a thermometer. To further validate it, they tried different forms of thermometric liquids. Water, paraffin, alcohol and eventually, mercury. As a result, the liquid particles inside the thermometer changed with reference to the movement of the particles in the thermometer's immediate surroundings. Mercury just happened to respond with more accuracy compared to the other thermometric fluids.

The disparate behaviour between the two entities (the thermometer and its surrounding) captures a substantive degree of coordination, signalling a particular room temperature. Communication, then, is achieved through the coordination of behaviour. For humans, we have been able to coordinate our behaviour to the point of moving our laryngeal muscles with such skill and acuity as to develop understanding from what another fellow human being says. It is a form of coordinated behaviour because laryngeal muscles move in a dynamically restricted way that influences the vibration of air particles; but air vibration will not make sense if the other human being has not coordinated itself to pick the bombardment of soundwaves it receives. Only when the recipient coordinates his auditory organs will it understand what the bombarding vibrations convey. Every movement will then be unique – if not the movement of the chest cavity, then the laryngeal muscles, if not that then the air coming out, if not that then the interaction between the air released and the surrounding air, if not that then the movement of the auditory organs, and even among these organs, there must be sufficient coordination to transduce the vibration with minimal changes to the intended message being conveyed. There is the potential for the sentence to be very different in this long, windy channel by the time it gets to the recipient. Language, however, still remains largely coherent and stable amid all this chaos. Similarly, a long journey is followed down the visual pathway when someone views written or typed words. Thus, every sentence constructed is anatomically and behaviourally unique. Coordination is, therefore, essential in all contexts for there to be communication.

With a little imagination, coordination of behaviour as the mechanism of communication could also apply to particles. There must be substantively well-coordinated behaviour that is precise enough for potassium to interact with chlorine. Such behaviour allows an electron to move from the potassium to the chlorine atom. This pattern of behaviour, similarly, is always unique, despite the ease and ubiquity of the reaction in different chemistry laboratories. Particles have therefore been able to formulate their own consistent pattern of behaviour, which is acknowledged as universal. This also shifts our understanding of how bacteria operate – they communicate through gene transfer among other strategies, which could mean all communicating bacteria are a unified organism. Following the same logic, the universal behaviour of particles could then imply that in our observable and shared universe, there exists a single organism comprising all the known and unknown particles. The universe I refer to is the one that scientists have always identified as our *shared* universe. I will further discuss the implications of such an interpretation in the next chapter. For now, let us go back to what you are reading at this moment – sentences.

To this point, we have seen that sentence construction demands well-coordinated behaviour and an understanding of what the sentence conveys also demands coordination of behaviour by the recipient. Two or more people communicating is an example of self-organization manifest. Let us break it down into small discrete steps. Speaking demands controlled release of air from the lungs, acute coordination of the vocal cords, specified moulding of one's tongue and associated movement of the structures of the oral and nasal cavity to produce a coherent sound vibration. In this entire process we have to remember that, as far as sound vibration includes particles that follow sound waves, these are organisms. On the other end, the recipient of these vibrations then makes sense out of the unique set of vibrations. It takes a bold move for an organism to coordinate its models of its reality in order to make meaning of what it receives. In this sense, there is some degree of anticipation if it is to formulate a model prior to receiving this unique set of vibrations. Once it has made sense of its reality, it then self-organizes and produces another set of uniquely moulded sound vibrations to the initial communicator, who also has to decipher what its universe has thrown at it. The only way these two correspondents can confirm that they have understood each other is if they continue with the same reference to the initial set of vibrations. This unique manifestation that I call a bold strategy, others call language. It is a manifestation of how every organism is able to self-organize and develop organized meaning from its universe of disorder. This is a marvellous feat. When two people are talking to each other, something special is taking place. You are allowed to stand in awe.

It is all the more astounding how the same steps explain the emergence of stability I was referring to. From chaos emerges order. From the unique set of vibrations, an entire discipline is developed and devoted to understanding linguistics and its evolution. Pinker argues that through evolution, humans had to develop a module that would facilitate this coordinated behaviour with astonishing accuracy. From the initial diversity of behaviour, a dynamically stable one was developed (of note is that this is a contested argument in light of other theories of language and our growing understanding of how the brain works). A similar story marks the evolution of the thermometer and selection of mercury as the more accurate thermometric fluid. Diversity achieved stability yet again.

The arrival of language started with the first organism before it became a set of grammatical rules, before the behaviour became well-coordinated by other organisms. Language is therefore another type of merger. It is largely unappreciated in the hard sciences but nevertheless evident. It also gives us an idea about how the other mergers (types of atomic bonds, for instance) might have emerged. I am yet to come across a hypothesis that tries to explain the emergence of the attractive forces of nature. Organismal Selection offers a possible means, following from how language develops. From planets and stars in a galaxy down to atomic and subatomic particles, there is the evident movement of particles. Movement begets behaviour. Cosmically, one form of behaviour might be called the inverse square law of gravity, and sub-atomically, it might be called the spin of the particle. At different scales, the different movements define the different forms of organismal behaviour, which if coordinated, constitute some form of communication.

There is one other bit that I also found amazing – the character of sound waves. Sound waves are different from electromagnetic waves in various ways. However, they share similarities, odd similarities. Electromagnetic waves are continuous. There are no discrete breaks in a wave; only a continuous stream. However, somehow, we can see seven colours when light strikes a prism. Similarly, sound is a continuous wave, with no breaks. We are, however, able to distinguish discrete words from this continuous stream. I found that truly astounding – the degree of self-organization an organism can undergo to make discrete meaning out of a continuous form is a highlight of an organism's ability to make the most of its universe while avoiding annihilation. For the visual stimulus, this would be different for different creatures in the same way the sound stimulus would be different. Take the example of the mantis shrimp and bats.

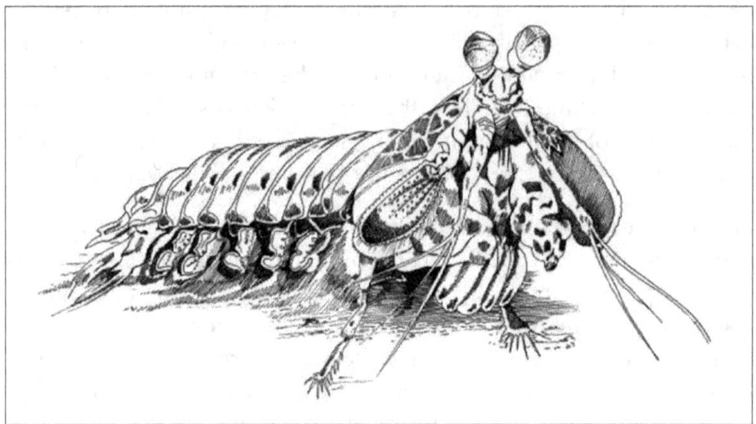

The Mantis Shrimp can scan its environment through different light wavelengths

The mantis shrimp has a unique set of eyes, held by independent, movable stalks. That however, is not its highlight. It has photoreceptor cells in the range of 12 to 16.[14] Humans only have three. The three allow us to view the seven colours of the rainbow, and, thus, can only see visible light. One cannot imagine what the mantis shrimp is capable of, as it can perceive wave ranges from deep ultraviolet to far-red and even polarised light. Furthermore, with such a broad range of options, it is capable of tuning from short to long wavelength ranges, a feature known as spectral tuning.[15] This is by far the most complex visual system known in the animal kingdom. The bats, on the other hand, produce ultrasound to appreciate their universe.[16] Since a sound wave has the property of bouncing when it hits a barrier, the reflection of the wave gives the bat an acute sense of where objects are in its surrounding. If the wave moves at a constant speed, once it bounces off a barrier and finds its way back, the bat can estimate its shape and distance. Light also has this same property, that allows us to view objects once they have bounced off them and onto out eyes. I remember loving how Toph Beifong, the legendary earth bender and my favourite character in the series *Legend of Aang*, was able to sense her surrounding using a similar property. This is a feat in self-organization or simply, complete marvel. Since the goal is always to avoid annihilation, self-organization is a trip in the history of the evolved traits of the multi-pigmented mantis shrimp and the sound-utilizing bat.

The Bat can scan its surrounding using soundwave echo-location.

Once more, it appears that this behaviour is evident in various organisms at different scales. There is a fractal pattern in behaviour, a self-similarity, at different organismal scales. But does this flattery extend to the particulate world? A deep dive shows that it is possible.

Particle Behaviour

Language is maintained through the coordination of behaviour. It is appreciable that in the long run, coordinated behaviour would then shape the organisms to be constrained by a particular set of behaviours in different situations. When you begin classes on how to drive a car, there are all these stimuli that you have to screen and beware of – the gear, the rear-view mirrors, the pedestrians, the lane, the traffic signs and others that everyone who drives a car for the first time can attest to. However, after regular practice, your behaviour becomes limited to a small set that you consider relevant. From an initial set of diverse behaviours, a stable behaviour was generated. It is just like laser light, where past a certain critical point, a single coherent light emerges from a chaotic set of light emitted by single particles. Coordinated behaviour is indicative of coherence.

It then allowed different organisms to get shaped into having a diverse or similar set of features. For human it allowed us to develop a larynx and a unique set of voice cords, with enough lung capacity for speech. It then allowed us to form specific areas in the brain that enabled us to coordinate the muscles of the respiratory system and generate well-coordinated movements to generate sound vibrations within the narrow sound ranges sufficient to create meaning for the recipient of such vibrations. The organism, therefore, got tightly coupled with its universe, and shaped it in the process. The reverse also applies. An exercise in trying out the Popperian schema can show how this coherence is attained.

As far as I can tell, coordinated behaviour can *only* be appreciated by motion. It is only when mother hen shrills that its chicks huddle under its wings. The chicks also start picking food only if the mother does so. The imitation game runs deep to the particles – through vibration. The single factor that links all particles is vibration. It is the first of two ways that I perceived OS could explain the kinetic theory of matter. The kinetic theory of matter states that particles are always in a state of continuous motion.[17] This is the definition given but it begs the question, why are they moving? Organismal Selection hints that by considering particles as organisms, they may be moving because that is how they coordinate their behaviour – that is how they communicate. It should not then come as a surprise that particles anywhere we have always known them to be, share this unique property. As previously stated, communication has a self-organization propensity, attaining stability in spite universal perturbations. As for vibrating particles, this behaviour has been present since particles appeared after the Big Bang. It has remained consistent billions of years later and I wager it will remain long after we are dead. This is self-organization at its best. It is order maintained for generations amid universal chaos.

Due to our scaled-up properties, we cannot unequivocally assert that particles communicate. We can, however, yield to the same logic of fractal behaviour across all scales of matter. If the aim of an organism is to communicate, there has to be a pattern or form of vibration of body parts that creates a wave with a signal to the recipient. For the signal to be interpreted, the vibration has to be matched to the model created by the recipient organism. This much can be appreciated through the behaviour and hence the level of coordination of the recipient. It bears little meaning if the recipient cannot decipher this behaviour. If the recipient has cracked the meaning of this vibratory movement, it can create another set of behavioural stunts hoping that the initial sender understands, and if possible, with reference to the initial message. This would then constitute language – coordinated behaviour with reference to a previous coordinated behaviour. For the particles, the all-cutting behaviour noted is vibration, which could account for particulate communication. This is the first way I interpreted that OS could explain why particles are constantly moving. I shall later elaborate on the kinetic theory of matter in relation to OS, in the next chapter.

There is more to this story than just the kinetic theory of matter. An example that I entertained was the phase transition of matter. Phase transition can be seen in the shifting of states of matter from either solid to liquid, liquid to gas or vice versa. Take for instance, propane. Propane is a gas, with various particles, and hence, various organisms. If we put all the propane gas molecules in a chamber, there is very little that happens besides random motion. Remember, this random motion is the particle wandering trying to find a means of avoiding its present state of continuous descent to certain annihilation. In such a state, the degree of coordination through particle vibration results in little communication with the other particles. They are autonomous and there is a very small degree of merging (as small as the weak gravitational attraction and van der Waals forces among them). When we start compressing the gas molecules in that chamber, the pressure increases for the particles to quickly find a means of surviving the impending threat. Under increasing pressure to find a solution, the degree of random motion is reduced since there is a continuous decrease in space for movement. Unexpectedly, their range of motion cohere and they start communicating, through coordinated movement. Again, self-organization results unexpectedly and unplanned among the involved organisms. We then have, at this scale, three mergers – coordinated movement, van der Waals forces and gravitational attraction – which is much greater (or more mercurial) than gravitational attraction alone. As a result, an emergent state arises – the propane gas turns into a liquid state. The newly developed merger has demonstrated the emergence of a new organism, with different properties. The reverse also applies. When the liquid is subjected to heat the emergent organism gets continuously destroyed and everyone seeks their own escape route as different particles. Phase transition from one state of matter to another demonstrates the emergence of new organisms. Emergence of new organisms happens when there is coherence – coordinated behaviour. Later, I hope to persuade you that the coherence shows an alignment of goals among the organisms involved.

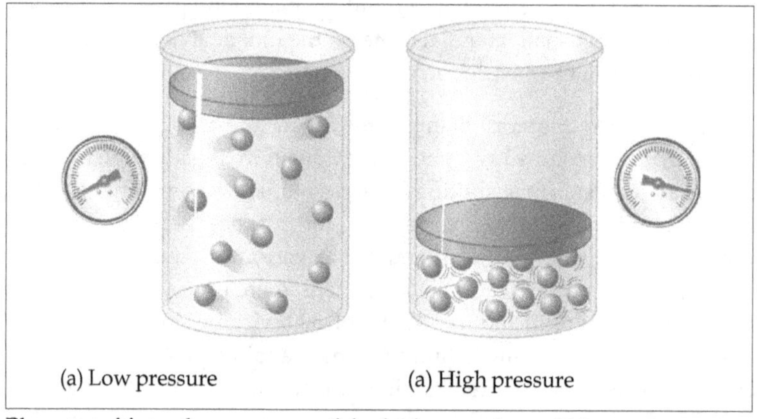

Phase transition – the gaseous particles in the container with low pressure can be change from the gaseous state to liquid state with increasing pressure as seen in the same container, now under high pressure, but only past a certain critical point.

If this is the case, it brings forth another type of merger that is evident across all scales of organisms. Chemistry has taught us about the various types of bonds; ionic, metallic, covalent, hydrogen, Van der Waals forces and possibly others that I might not be aware of. These are some of the mergers that have been verified constantly and independently in all chemical laboratories. There is, however, one other merger – that is, coordinated behaviour. When a child hears their parent talking we have few to no way of knowing whether they understand or not. As they continue to grow, even before they know how to speak, giving them orders is a way of coordinating their behaviour with that of their parents. We might not know if the child understands the words used in a stern warning, but the coordinated behaviour of a possible harsh tone and contorted face proves to be evolutionarily robust as to indicate that there is communication between them. It should then be easy to explain how when a baby meets another they immediately hit it off with their 'gibberish', which is their closest form of coordinated behaviour.

Judith Rich Harris showed how a child easily interacts with another, rather than an adult. The reason could be that the interaction is easy since they possess a range of behaviours that are relatively easily coordinated among themselves. The coordination of behaviour between a mother hen and its chicks could seems alien to a human being. That their behaviour is alien to the human being does not mean that there is no communication. There is coordinated behaviour between mother hen and its chicks. Borrowing the same logic, that particle vibration appears alien to our form of communication, does not mean that it is not happening. The vibration of particles can be deduced to be a form of communication between particles.

Across all scales, from the small particles to galaxies, it would then appear that there is a coordination of behaviour. This is a merger, a bond as different as the bonds already described. It is a form of self-organization, a property that equips organisms with the ability to escape annihilation. Jared Diamond explains the invaluable role the elderly play in a community.[18] They can be the difference between complete extinction or near survival of a community. From these arguments, it appears that communication, coordinated behaviour, is a special type of merger. It is crucial to the avoidance of annihilation like any other chemically acknowledged bond. Organismal Selection therefore not only explains the mechanism behind the kinetic theory of matter, but also breaks the barrier between particle physics, sociology, ethology, linguistics and any other field I might have left out. It almost makes one wonder why greetings are short, brief and often bear similar responses in return. Hello,

Hello; Hi, Hi; Ola, Ola; Good morning, Good morning. It might seem a bit of a stretch, but it all could be related to easy coordination of behaviour or imitation, or both. Fritjof Kapra, quoting Lynn Margulis, reiterates that chemicals do not organize in a random way, but in ordered, patterned ways. It is such similar ordered patterns that I believe result from coordinated behaviour.

In fact, I tend to think that coordinated behaviour can explain the various forms of mimicry recorded in biology. Batesian mimicry, for instance is a special phenomenon where a harmless organism (the mimic) mimics the conspicuous traits of another noxious type (the model).[19] Some organisms can change their outward appearance without having major shifts in their genotype, even though a major genetic mutation is conventionally accepted to be the first step to convert a non-mimic to a mimic.[20] It often leads to a lot of confusion even among experts when two totally different species bear similar physical characteristics. It is often the reason for the confusion between the Texas coral snake and a milk snake. These two snakes have similar patterns but the milk snake is harmless compared to the venomous Texas coral snake. There are cases where these traits are variably inheritable but, nonetheless, present. There have been instances where the model has gone extinct but the mimic remains existing, where the established pattern for the mimic remains.[21] These are genetic feats of self-organization that once activated (like Turing's autocatalytic phenomena) establish a long-term pattern with its predators (inhibitory phenomena).

Greyscale images of the venomous Texas Coral Snake (top-most) and the harmless two milk snakes (bottom two snakes).

The Immune System

Self-organization can then be considered as the property that aligns with the organism's goal of avoiding annihilation. One system that displays this feature throughout the life of any human being would have to be the immune system. However, before we appreciate how the system works, we would have to reconsider the commonly perceived primary function of immune system from defence to self-recognition. The immune system needs to first and foremost recognise self before it attacks anything that is not part of self. This has to be a dynamic form of self-recognition that allows for the exchange of the organism and substrates it identifies in its universe. A rigid or static immune system is not bound to survive. A dynamic form of self-recognition allows for small errors or small particles to merge with the organism. It would allow for the organism to ingest food, then attack it because it does not recognise such a large, hard, complex material that we commonly recognize as food. The human then retaliates by mechanically and chemically breaking down this organism from the mouth all the way to the intestines. Once it has broken the food to fine bits, integration with the human organism is then possible. The fine molecules bypass the detection of the human organism as foreign and can, thus, be allocated to various parts of the body. Digestion is then an exercise in the annihilation of an organism (food) by breaking it down into similar particles that can escape the dynamic, albeit vigilant, immune system. Its vigilance is noted by its high sensitivity to detecting non-self from self. For acute detection of non-self, an organism has to have a *constantly updated* version of self. Self-recognition should therefore come first to non-self-recognition. Defence would then be a secondary role.

It also means that the sensory systems should be just as dynamically shifting as that of the immune system. In a discussion with Steven Strogatz, Cornelia Bargmann confirms that the sensory system is arguably one of the most rapidly evolving systems of the body.[22] I would argue that the digestive system, the immune system and sensory system are all connected to the organism's goal. There must be an acute detection mechanism that identifies harmful from viable organisms (sensory and immune processes) and an efficient mechanism for eliminating the organism (mechanical and chemical digestion processes). An analysis of how immune cells operate does not distinguish them from the digestive system. For digestion to happen, there must be swallowing. Very many cells of the immune system use the same process, by swallowing small pieces of foreign material. Once swallowed, there are the mechanical and chemical processes of destruction. Immune cells chemically destroy the foreign material. The striking similarities between the two systems show an evolution towards detection of foreign organisms and a radically shifting and updating process of self-identification. Improvement of self-identification comes with the need to have an ever-improving sensory system. All these systems work in concert to enhance sensitivity to viable from non-viable organisms for merging. Their synergistic efforts are an example of self-organization at work.

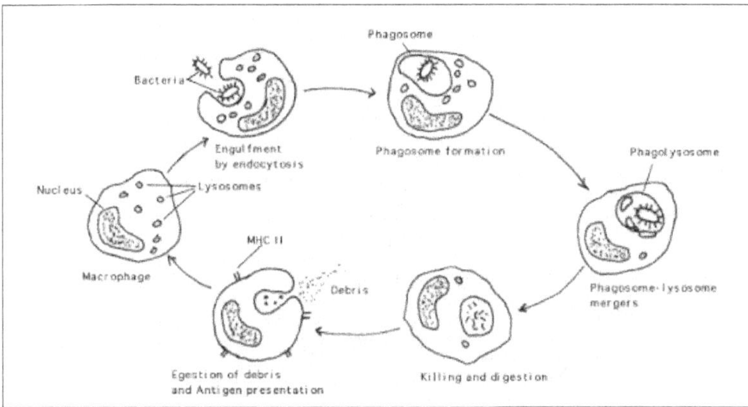

Images showing the similarity in cellular feeding and human feeding. In (a) a human being is taking a meal. In (b), an immune cell is also taking a 'meal'. These two processes are similar. .

What about the ingested food? Is it not an organism? Should it not also display resistance? It does. The ingested food displays resistance the moment the human being had to use energy to break it down first mechanically, then chemically. This is a bold strategy both from the food, with its own structural organization, and the human being. The digestive system is therefore part of the immune system. The fine organisms (finely digested material) can then merge with the large organism (animal or human being) to avoid annihilation.

Mergers are an evolutionarily stable strategy evident from the moment particles started existing down to the present day and they continue every time you take a meal. This dynamic system, comprising the digestive, sensory and immune systems, which allows for a fine set of organisms to merge with itself, demonstrates the human ability to self-organize without *significantly* destroying the self. The organism is able to break down another, say lettuce, while preserving a significant portion of its integrity, so that it does not lose its identity. Every attempt at preserving its identity is in line with the goal of avoiding annihilation.

To further clarify how the human being is structured to achieve this goal it is important to note that the immune system extends to all the known barrier mechanisms of the body. All epithelial linings are barrier mechanisms. They are seen in different parts of the body including the digestive system. The skin is also a barrier. The brain has a barrier. The eye has a barrier. The gonads have barriers. And in most of these barriers there are receptors. That is to say, the sensory system will inescapably be connected to these barrier mechanisms, and they are. The integrated efforts of these seemingly disparate but connected systems aid in the preservation of the organism's self-identity through the elimination of or the prevention from perceived foreign organisms. Self-identity becomes a primary concern while the destruction of foreign organisms becomes secondary. If recognition of self becomes an issue, auto-immune diseases crop up.

There is substantive evidence to support our assertion that the immune system serves primarily as a system that recognises self before none-self. Classically, the cells of the immune system have always been described with military metaphors such as the 'foot soldiers of the body', 'front-line soldiers' or the 'guardians'. However, these metaphors were only created after studies confirmed viewing the antibodies in action in the presence of foreign material. Even the name itself, 'antibody', shows that there must be an understanding of what the body is first before targeting those that are 'anti' – those that are not of the body. Under normal circumstances, it has been found that these military cells are bound to virtually all types of cells of the body. Did they all of a sudden turn 'soft' or were they 'soft' from the start? The idea of binding to cells even in the absence of a foreign body possibly shows that the system is initially concerned with having an updated version of what the self is. Moreover, various immune cells are constantly traversing the body in 24-hour traveling trains known as the lymphatic channels, which traverse every organ in the body.[23] Organismal Selection explains why this is so. Constant imbibing and interaction of an organism with its universe exposes it to a constant change of what the organism is – the organism has a plastic dance in an arena where it has to preserve its organization by organizing itself – self-organization. Capra's *The Web of Life* highlights that this is what Francisco Varella and the immunologist Antonio Coutinho found out:

> The mutual dance between the immune system and the body ... allows the body to have a plastic identity throughout its life and its multiple encounters'.

I further insist that even the digestive and sensory systems are involved. In fact, all the systems are involved, in a coordinated effort in preserving the individual.

The dynamic dance of the immune system can allow some particles to co-exist with the human being or your organism of choice, and, hence negotiate mergers. But we had already clarified that a merger results in the formation of a new organism. A human being, therefore, is not just a human being because of his genotype, but exists due to the numerous bacteria that flood his body and line epithelial barriers. Gut bacterial flora, for instance, have been noted to alter the mood and affect of human beings, constituting some physical evidence of 'gut feeling'.[24] They play a very important role in immunity as the epithelial barriers. A relationship, if formed, between the different colonies of bacteria at different parts of the epithelial barriers, create hierarchical systems. The purpose of the bacteria then serves the epithelial barriers and the purpose of the epithelial barriers serves the bacteria. This is an example of a *stable* hierarchical relationship, where the purpose of one serves the other and vice versa. It incorporates both the autonomy and dependence of organisms involved in a merger. The tight coupling of the hierarchies is the reason organisms exist as they do. A bear, a rat, a snake, a crocodile, a Venus fly-trap, a mosquito – all contain different parts consisting various sub-systems of organisms but do not attack itself. Such an offensive strategy would break the hierarchical relationship and end up destroying the merger that is supposed to serve every organism involved.

New mergers result in the formation of new organisms, only when the vigilant system comprising the immune, sensory and digestive systems, allows it to happen. The fine material, in the name of nutrients that can be utilised by the human being, establishes a relationship where both organisms benefit from the merger. The broken-down protein or carbohydrate or micronutrient can form a stable hierarchical relationship and, hence, get accepted in forming something much greater than themselves. The vigilant system, then, distinguishes the purpose of hierarchies – the purpose of the higher ones serves the lower ones and the purpose of the lower ones serves the higher ones. The purpose of nuclear material serves the cell and the cell serves the nuclear material. The purpose of the cell serves the tissue and the tissue serves the cell. The purpose of the tissue serves the organ and the organ serves the tissue. The purpose of the organ serves the organ system, and the organ system serves the organs. The purpose of the individual serves the organ systems and the systems serve the individual. The powerful role of hierarchies helps explain why genotypes influence phenotypes. It is not as straight-forward as the map of genotype-phenotype makes us believe.

Hierarchical relationships are much more complex than the simplified two-way order I have described. The simplified version is in the interest of showing how a simple relationship can generate a stable complex system. For instance, an attempt at demarcating hierarchies past the cellular level has been tested and shown to obey simple rules related to gradient steepness and particular concentrations of materials. These regions have the special feature of self-organization that can be well described by both reaction-diffusion systems and positional information systems. These special and sensitive regions of the body are known as morphogenetic fields. This idea of morphogenetic fields took a back seat following the discovery of the DNA. However, its role in embryological development is indispensable.[25]

However, hierarchies extend past morphogenetic fields. We can therefore extrapolate the dynamic relationship that hierarchies manifest. The purpose of the parents serves the child and that of the child serves the parents. The purpose of the family serves the community and that of community serves the family. These hierarchical relationships help illustrate how an organism is not just what biology tells us but an emergent entity from mergers of various types. As earlier explained, one need only introduce a credible threat to appreciate the presence of an organism. For the tightly coupled stable hierarchies and their multiple combinations, their existence is seen by subjecting each of them to some form of credible threat. There is bound to be resistance, in line with self-organization. The hierarchical systems display the autonomy as well as the dependence of organisms.

As we come to the end of this chapter, the powerful role of self-organization should be evident to the reader. Key properties of self-organization to remember are:

- Coherence, which signifies the existence of an emergent entity or system and its coordinated activities.
- Power laws, which is the tendency of systems to resist multiple perturbations in an effort to preserve their coherence.
- Hierarchies which captures the autonomy and dependence of a system.
- Resilience which is the capacity of a system to self-organize.

It should also be clear how self-organization, which although is tied to the goal of avoiding annihilation, manifests several specific outcomes spontaneously for reasons unknown to us. This happens at various hierarchies and contributes to the resilience of the organism. A *tendency* to avoid annihilation through the property of power laws then equips all organisms with at the very least, baseline cognitive properties. Particles, which we have labelled as organisms, therefore, appear to have cognitive properties such as that of self- identity or some sense of self-awareness. A much deeper understanding of self-organization shows that it could be the gateway to building an understanding of self-identification and self-awareness. Just as the self-organizing immune system is in a constant state of self-awareness by flooding almost every cell in the body, it helps to bear in mind the identity of the organism. This self-awareness then shapes the organism's identity – it brings forth self-identity. Self-identity, as a cognitive process, would then constitute the organism's consciousness. Thus, consciousness becomes a downstream function of self-organization.

Consciousness can then be described as a state of being aware of oneself to the point of organizing oneself to avoid annihilation. Awareness of a harsh or a pleasing stimulus is needed to initiate mechanisms to mitigate or eliminate harmful stimuli or attract or incline towards pleasant stimuli. Such cognitive processes stem from the tendency of an organism to avoid annihilation. First by having a *sense* of what it is, and secondly, by resisting efforts to annihilate it by forces external to itself. That is, firstly, by self-identification then by self-awareness of one's surroundings. Self-identification and the tendency to preserve oneself from annihilation produces self-organization. What this would mean is that cognitive processes, such as self-identity and self-awareness, are fundamental to any particle that has ever existed.

These processes are only likely to have evolved with more mergers to the point of emergence of what we identify as consciousness. In the simplified jargon of calculus, the local rate change generated by the particle mergers results in a cumulative change of self-identity known as consciousness. It might appear that for a human being that regularly feeds, the consciousness should be shifting. It does. Feelings such as satiety, reduced hunger with every bite or sip, and influence of mood by gut flora all contribute to altered conscious states. This does not eliminate the overall state of self-identity. The local rate change is small in comparison to the cumulative change that brought forth consciousness as a marker of self-identity. The removal of an electron from the sodium atom changes it significantly, to an ion, since the electron is, relatively speaking, a significant portion of the sodium atom. But the removal of the sodium atom in an organism such as you and me is negligible compared to our relatively colossal mass. Similarly, the local rate mergers of fine particles have few effects in radically shifting the consciousness of the organism, which constitutes the cumulatively preserved, albeit gradually changing, consciousness.

I have always found it amusing how the question of self-awareness is responded to in the positive, only when asked, but one continues about their duties if not asked such a question. When you ask your friend: Are you aware of what you're doing? They would often reply 'yes', unless they are about to perform a crazy stunt. When knee-deep into a problem, in life, a laboratory, or even in an exam room, one tends to get immersed so much in the problem that they lose a sense of self-awareness. One does not become conscious of their position, posture, shifting glances, but remains immersed and lost in thought. We are constantly in a state of sub-consciousness but conscious awareness often if not only emerges when asked if we are aware of our surrounding or aware of ourselves – if we are self-aware. As I have reiterated before, the identification of an organism is evident if the organism is subjected to a credible form of threat. Asking such a question is similar to posing a threat to an organism. It is threatened by the idea that someone else thinks they are not existent or aware of itself, as an organism. Thus, the question is always if not often answered in the positive. Such a question is no different from a harsh universe attempting to annihilate an organism. The organism would then assert a bold answer – 'Yes, I am aware of myself', or, 'I am aware of what I am doing'. Bearing this suggestion in mind, it then follows that consciousness could have evolved as a cognitive process to defend an organism from the threat of its universe. The platinum proof thus far of testing existence of an organism, an entity literally bigger than the sum of its parts, is by subjecting it to a credible threat. The fractal nature of organisms at all scales – of coordination of behaviour, of different types of mergers at different scales or in general, of self-organization – might appear indeed as the bold steps organisms take to avoid annihilation in their respective universes.

Before we jump onto another chapter, I have to highlight an aspect that might elude yet prove useful to the reader in understanding the role of mergers in relation to self-organization. A simple way to look at mergers is the resultant emergence of a new organism. However, through self-organization, mergers result in much more. We have previously shown how mergers reduce the overall sample sizes of organisms. The other thing that it does is reduce the degree of freedom of organisms. When a husband and a wife get married, the degree of freedom for each individual in the relationship gets reduced. The merger results in various restrictions. The husband might have to now consider two people when making the monthly budget and the same case applies to the wife. Similarly, when a hydrogen nucleus merges with another nucleus, the range of options for each reduces. From a space of large possibilities, mergers reduce the range of options for organisms. Thus, self-organization increases predictability by significantly reducing the space of possibilities for the constituent organisms. When players get onto a soccer pitch, each player's role is constrained to that of the team. He or she cannot just run around aimlessly or create his own rules. One cannot just hold the ball and run all the way to the opposite goal. Outside the team, it is individually possible. However, when a game is on, it is not. The range of options for the player is reduced by being merged with the other players in the team. Self-organization has this unique ability to increase the predictability of a system by reducing the space of possibilities for its constituent organisms.

Combining the three S's, statistics, symbiosis (symbiogenesis) and self-organization, we can then see how we have transitioned from the organism as an entity with untold variability, to one with a reduced range of variability. Formally, through statistics, the organism remains the fundamental unit of diversity. Empirically, through mergers and self-organization, there are more robust and testable states. We might be unable to detect where a proton or an electron is with accuracy, but when the two merge to form an atom, detection increases due to their reduced degree of freedom of the emergent organism. Nevertheless, strictly speaking, it results in more qualitative than quantitative prediction. It would explain why 'animate' and 'inanimate' sciences have precise descriptions about various systems and yet know very little about their future outcomes.

Self-organization captures the third 'S' which explains the arrival of the fittest. Even though I am blown away by the first three S's, it is the last one that I initially found intriguing. The first S, statistics, was the first clue that the theory is different from Natural Selection. It was when studying aspects of Game Theory, of cooperative and defective strategies, that I concluded that the first organism neither had anyone to cooperate with nor compete against. It opened doors to its exploration and its beautiful conclusions as I have hopefully tried to explain. Symbiogenesis was an idea I contemplated through the avenue of statistics. That is, how mergers end up increasing the diversity potential in a universe. I did not narrow it to just the cellular organisms but included it as a process also evident in particulate organisms such as atoms and quarks. The general term 'merger' ought to espouse the idea of symbiogenesis. This led me to self-organization, which was needed for there to be successful mergers between two organisms. Self-organization was therefore a discovery I made much later. Much like the order it creates, self-organization was a spontaneous idea.

It has been suggested by Stuart Kauffman that there is more to evolution than just Darwinian selection. Darwinian selection speaks of a gradual progress, but self-organization describes spontaneous order – nothing gradual. Together with other scientists, Kauffman seeks to develop general laws of self-organization, enough to write a book on them. A single chapter on the topic hardly captures the progress done in this field. Their evidence, nevertheless, is in keeping with our idea of how self-organization precedes selection (which we have operationally called annihilation). But even before self-organization, I had developed the fourth 'S' – size. Let us explore it together. If the first three S's haven't convinced or astounded you, size will definitely knock your socks off.

CHAPTER SIX

Size

*It's five dimensions, six senses, seven
firmaments of heaven to hell,*

Eight million stories to tell,

*Nine planets faithfully keep in orbit with
the probable tenth,*

The universe expands in length

– Mathematics, Mos Def

The map I have highlighted throughout this book reminds me of numbers. In mathematics, there are rational numbers, which can be expressed as ratios, particularly as a ratio of an integer to a non-zero integer, hence the name. Those that cannot be expressed as a ratio are known as irrational numbers. There is a special group of irrational numbers known as transcendental numbers. All transcendental numbers are irrational but not all irrational numbers are transcendental. Examples of transcendental numbers are Euler's number (e) and Pi (π) which are the largest of numbers known to mathematicians, hence the name transcendental. We do not know where or how Pi ends or even if it does. It is just a continuous series of numbers with no known repeated sequence. What is surprising is how in a sea of transcendental numbers we have the ability to identify the small, almost negligible islands of rational numbers, ordinal enough for us to develop sequences. There is a high degree of order in these small islands. They remind me of an organism localised and its widespread transcendental universe. More than that, the parallel modes of comparison remind me of size. Size wills an organism much more than just the face value of the term. As Mos Def reiterates and as we will find out, it's all mathematics.

Size, as the next 'S' expounding on the arrival of the fittest, is the last component that is tightly linked to self-organization and symbiogenesis. One strategy that illustrates this link is cooperation. As an evolutionary strategy, cooperation is viewed as a stable and reliable strategy – bold enough to increase an organism's chance of survival.[1] While it might appear that cooperation is intentional, it might also just be an emergent strategy following the alignment of goals of autonomous organisms. Cooperation, in the broad sense, is a type of merger following overlapping of different organismal goals. This usually culminates in the emergence of another organism in the same way a new organism emerges when two atomic particles merge to form a molecular compound.

The example of diffusion, essential in bacterial survival, might appear simple but it captures a scaled-down version of the emergence of and divergence from cooperation, which, to my knowledge, is present in all biological systems.

A simple bacterium can ingest and release waste material by simple diffusion.[2] Here we have two organisms – the bacterium and the particulate waste products. The goal of each is to avoid annihilation. The recognition of this is seen in the ability of the bacterium to contain a certain amount of 'waste material' before it eliminates it via diffusion. Viewed in this way, they are not really waste material but a sub-set that makes up the emergent organism that is the bacterium. However, as the 'waste material' continues to accumulate, the size increases. The bacterium, with time, cannot allow itself to stay with the waste material since it poses a credible threat to the bacterium's life. Credible threat to any organism is met with resistance. The waste material cannot stay inside the cell if it is to exist with the bacterium. The two goals create a conflict and the merger has to cease. Due to this discordant alignment of goals, self-organization at this level allows for the particles to exit the cell. This happens through the process of diffusion. There was some degree of cooperation when the quantity was small but once the size started increasing, it posed a threat to the different hierarchy-level organism, that is, the bacterium. From how hierarchies work, the purpose of the lower ones is to serve the upper ones and the upper ones to serve the lower ones. In this case, the low-level hierarchy was not serving the high-level one and the high-level one therefore decided it would not serve the low-level hierarchy. The cooperation was destroyed and diffusion happened. These particles are only considered 'waste' with reference to the bacterium but they are nonetheless organisms who continue seeking mergers with other particles to postpone their annihilation through atomic or molecular versions of symbiogenesis. Self-organization, symbiogenesis, and size are thus tightly linked.

Now, when there are more cells together, such as in a tree shrew, simple diffusion is not an efficient method. A more co-ordinated approach is needed. Self-organization would then lead to the development of tube systems to facilitate this pressing need. Even then, diffusion still plays a role, but often at the terminal ends where there is exchange of particles. An example of a terminal end is the terminal bronchiole in the lung, where oxygen diffuses into the bloodstream and carbon dioxide diffuses into the alveolar space of the lung to be breathed out. The optimised architecture of tubes leads up to the terminal ends to facilitate the exchange of materials. In bacteria, these tubal systems are not needed. In a tree shrew, they are. The goal presses each organism to self-organize in order to eliminate 'waste'. In the tree shrew, it does it while preserving the processes that allowed it to do so when it was small – as small as the zygote stage when the shrew was a single-cell organism immediately after fertilisation. Helpful processes will be preserved as the organism increases in size. Therefore, some elements of diffusion remain even in large organisms. Size and self-organization, again, remain linked. A tendency to avoid annihilation determines the organization of the organism. These two examples, the unicellular bacterium and the tree shrew, intuit the next 'S' – size. This, however, is not how I arrived at the notion of organisms tending to increase size.

Size and Relativity

I first encountered the idea of size when reading Albert Einstein's book, *Relativity*, and was intensely fascinated by how matter distorts space and time.[3] The larger the mass of the object, the more it distorted space and time. Even more astounding was the idea that for a moving object time moved much slower than for an object at rest. It was from Steven Hawking and Leonard Mlodinow's *The Grand Design*, that I got a clearer picture of how relativity works; specifically, how matter distorts space and time.[4] I was able to understand the principle of equivalence, which broadly states that there is no way of determining if an object is in uniform motion, or if it is stationary in a uniform gravitational field. To appreciate this principle, here is a thought experiment.

Imagine you are in a lift in space, floating freely. This lift is attached to a cord at the top. Now imagine a being pulling the cord uniformly, so that the lift starts to accelerate. Automatically, there appears to be a pull on your legs. It can be regarded as the 'floor' since if you were to drop a pencil, it would fall to the 'floor'. If you had a pencil and a feather, both would fall to the 'floor' at the same rate since you are in space since there is no friction in space. But since you are inside the lift, you can never know if you are being pulled by some celestial being or if you are experiencing life as we normally do on Earth, a planet with a fairly uniform gravitational field. The mass (pull) that is developed due to the uniform motion of the lift is the same as the acceleration of the lift. Since gravity is a force that forms from acceleration (hence gravitational acceleration), the inertial mass becomes equivalent to the gravitational mass. Similarly, if one were sitting on a train that did not produce any sound when moving, one could not tell if the train is moving or not. Hence the principle – one cannot detect if they are in uniform acceleration or at rest in a uniform gravitational field.

Using this principle, one can imagine a spaceship as long as the speed of light in one second (3×10^8 meters long) in uniform motion. If a sequence of light signals is emitted at the tail-end of spaceship, one second apart, the recipient of the signal on the nose-end will not receive the signal at

141

the one-second intervals they are sent. Since the spaceship is in uniform motion, accelerating, the signal will continue delaying more and more. Also, the more the acceleration, the more the signal is delayed, and the slower time appears to be moving. If we were to test the principle of equivalence, it would then predict that if the same ship were stationary, on a launching pad on Earth, the signals sent from the nose-end to the tail-end would be received later and later with every signal sent, since the ship rests on a fairly uniform gravitational field. This would only happen if a body in uniform acceleration (such as our accelerating ship) is equivalent to a body at rest in a uniform gravitational field (such as our ship on Earth). It has been shown that if the two individuals had an atomic clock, they would record different times. The one closer to Earth would move much slower since it is closer to Earth's gravitational field than the one at the nose-end of the ship. If we were to go back to our space-ship moving in uniform motion, a simplified consequence that we can therefore get from this principle is that time moves fast for stationary objects and slower for objects in motion.

Using the principle of equivalence, a body in uniform motion would be no different from a stationary object in a uniform gravitational field. Large objects contain large gravitational force compared to smaller objects, and for them, the pace of their life would be slower than for smaller objects. To reduce the pace of life, the object has to be large. If it is large, it will have a stronger gravitational field and, hence, a slower pace of life. Furthermore, since acceleration is nonlinear (m/s^2 – meters per square second) compared to velocity, which is linear (m/s – meters per second), for every second interval, there is a square difference. Therefore, larger objects would tend to always get much more time than smaller ones or, rather, the pace of life was much slower in larger objects than smaller ones in a nonlinear scale. You tend to get a non-linear profit for an increase in size – some kind of economies of scale. The conclusion I, therefore, reached was that for an organism to evolve, it has to at least consider increasing in size, so as to delay its annihilation by reducing its pace of life. I called this bit of my theory, the 'Special Theory of Evolution', seeing that I explored the idea following Einstein's theory of relativity.

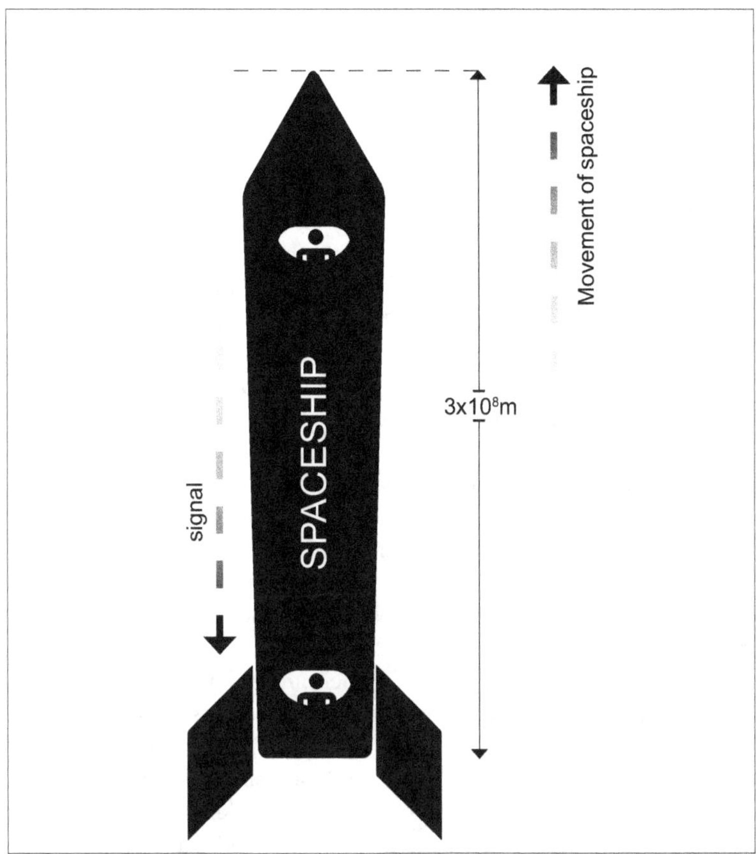

A spaceship, in space and in motion. The person at the head of the space-ship sends a signal from the tip of the head of the spaceship to be received on at the tail-end by the other astronaut. The spaceship is $3×10^8$ metres long, the equivalent of the length of light captured in one second. If the spaceship is stagnant, and the signals are sent after every second, the signals will be received after every second. However, for a spaceship that is accelerating, for instance, at $5m/s^2$, then the signals will progressively delay, each taking longer than the previous one.

I had already read Jared Diamond's book, *The Third Chimpanzee*, and knew that there were other animals such as the turtle that got to live for long in spite of their size. I could not dispute such facts nor how the naked mole rat could live for such a long time, in spite of its similar size to other rat species that die way earlier.[5, 6] I, however, did not consider that as falsifying evidence, rather that there are various ways of avoiding annihilation. I entertained the thought that if other factors were levelled, then size would determine which ones lasted longer and which ones did not. At this point, I had not yet considered that particles were organisms. Neither was I aware of the Kleiber's ¾ law. So, I went out to test my theory, and correctly enough, I found that larger animals lived longer than smaller ones. A fruit fly would die earlier than a house fly, and a house fly earlier than a cockroach, and a cockroach earlier than a rat, and a rat earlier than a cat, and a cat earlier than a dog, and a dog earlier than a horse, donkey or zebra, and these much earlier than an elephant. Among the whales, I even discovered that the life span of a blue whale was long but that it differed from that of other whales that almost weigh as much as it does. Whales, naked mole rats, turtles and tortoises initially turned out to be the outliers. Nevertheless, I still believed that these complex adaptive systems had a tendency to increase in size and there could or should be reasons for their anomalous lifespan. It turns out that there were reasons that still proved to be size-related. Until then, I concluded that despite my difficulty in developing a model that could accurately determine this, a foundation such as that of relativity was enough grounds to conclude that size does delay death. The non-linear property of acceleration significantly showed that size did have a role in avoiding annihilation.

My first assertion was that for an organism to have an edge in evolution, it had to at least consider increasing in size. This was matched by all organisms, particulate or otherwise. For a particle to avoid annihilation, it had to merge with another in a form of particulate symbiogenesis. This would result in not only a new organism but also an 'increase' in size. The particle would not increase in size, but the merger would result in an organism larger than the particle and hence one that would have a slower pace of life. Thus, a double reward stemming from the merger.

This process is replicated in every laboratory when particles fuse with other particles to form various molecules and compounds.[7] As for cellular organisms, if there was a single-celled organism, it had to evolve in a manner as to increase its size. This conclusion further agreed with the development of the prokaryote to the much larger eukaryote and ultimately to the multi-cellular organisms. Even the mechanism of symbiogenesis that allows for eukaryotes to ingest other smaller bacteria is a mechanism of size increase.[8] Slime, which can exist as individual cells, usually merges into a single, multicellular organism in the face of reduced food resources. They then track means of seeking food by intelligently exploring their surroundings.[9] There are instances where they also form stalks.[10] Cells that form this stalk run the chance of not surviving compared to the others that form other parts of the multicellular slime. Nevertheless, they still do this in an astonishing feat of coordination. As a single-celled organism it would not be able to do this.

While such examples might hint at altruism, it would also gain another interpretation, a hierarchical interpretation. That is, an emergent organism forms after the cells merge with the higher hierarchy serving the lower ones and the lower ones serving the higher ones. Furthermore, these cells inclined towards an increase in size. An increase in size would lead to delayed annihilation. Symbiogenesis and self-organization would, more often than not, result in an increase in size. I am yet to encounter a situation where there is symbiogenesis and without some element of self-organization and a consequent increase in size.

Before we explore the other size-related factors, I would want to explain further the other aspect I extrapolated from relativity. The first was the idea of space and time. The universe is expanding, so space is continuously increasing. In addition, time, on average, tends to move forward, so that is also shifting. The coordinates that an organism would occupy would therefore be unique to that organism at any particular time in the history of our shared, observable universe. I called this concept spatio-temporal niche – the space and time that a single organism can only occupy by itself. No other organism can occupy this niche. I found that this new way of looking at the ecological idea of niche was another way of asserting that an organism was the fundamental unit of diversity. The map would also stand out, as every single organism would have its own unique spatio-temporal niche. The consistency of OS was still intact and aligned with relativity.

The other one had to do with delaying annihilation. For bodies to slow time down, they can try the option of increasing in size. The other option is for it to be in a state of continuous motion. While large objects can appear to be motionless, their size reduces the pace of their lives based on the principle of equivalence. On the other hand, small particles can compensate for this is by being in a state of continuous motion. The theory of relativity asserts that an object in motion slows down time. As organisms, such a move can be considered an evolutionarily stable strategy for organisms the size of small particles. This would then be the second reason to explain the kinetic theory of matter. The theory states that matter comprises particles that are in a continuous state of motion. Even in objects as large as Jupiter, which might appear relatively motionless, our understanding of hierarchy reminds us that at its atomic level there are particulate organisms in continuous motion. Hence, OS deduces that there is no particle, and hence no organism that is not in a state of constant motion.

I have discussed vibration or continuous motion, as a means through which coordination of behaviour facilitates communication among particles. From the precepts of relativity continuous motion also reduced the chances of a particle being annihilated. That is to say that by being in continuous motion the organism slows down time. The principle of equivalence can still be called upon to arrive at this conclusion. To remind ourselves, the principle states that there is no way of telling if an object is in a uniform state of motion or at rest in a uniform gravitational field. If we consider slowing down time as a strategy, increasing in size (or more specifically, mass) should be equivalent to being in a uniform state of motion. Since a small particle has already lost in the game of increasing size and, hence, a strong gravitational field, it considers an equivalent strategy – increasing its state of motion. Organismal Selection then, through the principles of relativity, explains why the kinetic theory of matter is as it is. It does not just *describe* the state but explains *why* it behaves that way. Relativity thus helped me understand the kinetic theory of matter and validated my assertion of particles as organisms tending towards delaying annihilation through maintaining a continuous state of motion. Perhaps it is relevant for the reader at this point to remind themselves of the Popperian schema that highlights the boldness and sub-optimality of each of these options – that of creating mergers and increasing in size.

Since small particles are in a constant state of motion following the kinetic theory of matter, I figured that it could also explain the simple processes of diffusion and even osmosis. For diffusion, it incorporates various aspects. Firstly, by being in continuous movement they are reducing the pace of life and extending the length of time they will be in existence – they slow down time. Secondly, their inescapable inclination to their eventual doom creates situations that allow for them to be risk-seeking. They would rather gamble by taking a chance at the low probability of a big gain compared to the near-certain descent into annihilation. Hence, unreactive molecules in one compartment cannot do much for themselves by staying in a single position. Since they do not chemically react, they cannot fuse with each other again by forming stronger mergers. In such a situation, their annihilation can be considered to be fast approaching. They would rather take the low probability gamble and seek other conjugations with other particles by dispersing from regions of high to low concentration. It would do them a double service first by being in motion, as a continuous state of motion delays their pace of existence. Secondly, they hope they would bump into another organism that they would get into a stronger merger with. Thus 'passive' diffusion *might* not be passive after all.

Take the example of hydrogen gas on Earth long before mammals and other large living organisms roamed the Earth. The different hydrogen molecules seized the opportunity to diffuse and search for a fusion stronger than the mere inverse rule of attraction of gravity and van der Waals forces. Some merged with carbon to form methane (CH_4), others with sulphur to form hydrogen sulphide (H_2S) and others with nitrogen to form ammonia (NH_3). Luckily, some particles encountered oxygen and formed water (H_2O). It is likely that the water molecule then became the first organism of its kind formed by the first hydrogen molecule that fused with oxygen. A hydrogen bond, which is stronger than gravity and Van der Waals forces, also emerged. This happened on the surface of early planet Earth. In other planets, to our knowledge, there was no such fusion, so the particle continued to move more and more – this is an example of the low probability gamble that lies in such situations. Such gambles hardly paid off. But for those on Earth, it did.

Similar reasons can apply to osmosis. Solvent particles would seek other forms of conjugation or mergers besides gravity or the constant force that holds them together. The low probability gamble even involves crossing a semi-permeable barrier, adding to the overall tendency of particles in their respective universes being inclined to annihilation. Once they have merged with the solutes on the other end, they have formed a new merger and a resultant new organism. They form a much stronger bond with solute particles than the bond that held the solvent particles together. Using the fourfold pattern of preferences, it yet again appears that behaviourally, particles are no different from humans or for that matter, other organisms.

Diffusion is a relatively simple process compared to reaction-diffusion. From the previous chapter, we highlighted that reaction-diffusion (RD) was one of the suggested mechanisms for self-organization. Besides the preceding word 'reaction', the process nevertheless, involve diffusion. Reaction-diffusion appears active since it involves spontaneous localised activation of some particles and widespread inhibition from others. This is so since RD experiments demand that there be two particles (morphogens) interacting with each other. But the process still involves diffusion. In the simple process of diffusion, there might appear to be no other particle (morphogen) but the particle movement is with the same intention of 'reacting' with another, just like those seen in RD models. Both are forms of diffusion with particles taking the low probability gamble of interacting with another particle. Thus, diffusion and reaction-diffusion are, respectively, mechanical and chemical processes that particles resort to minimise their chances of annihilation. Understanding why particles diffuse could help understand why RD models behave as they do (Additionally, should the diffusion also not interact with surrounding particles, it can generate a diffusion gradient which can used to define positional information (PI), another concept we encountered in the previous chapter. This is the other model, besides RD that can be used to explain self-organization. Organismal Selection therefore further explain, at least partly, why PI and RD are successful morphogenetic models).

Even more baffling is considering that the same principles could explain the statistical interpretation of entropy developed by Ludwig Boltzmann.[11, 12] Boltzmann developed a simple way of describing entropy using combinations of positional states of particles in an enclosed system. These combinations would emerge by natural processes such as diffusion. Picture a sealed, transparent cuboid divided into two compartments by a cardboard placed in the middle. Eight gas particles would then be introduced in only one of the two compartments. The cardboard has a hole that is just big enough to allow particles to move to the other empty compartment. The position occupied by the particles can then be calculated using the formula for combinations:

Combinations = **n!/(n-r)!r!**

Where **n** = 8, and **r** = the number of particles that get into the other compartment.

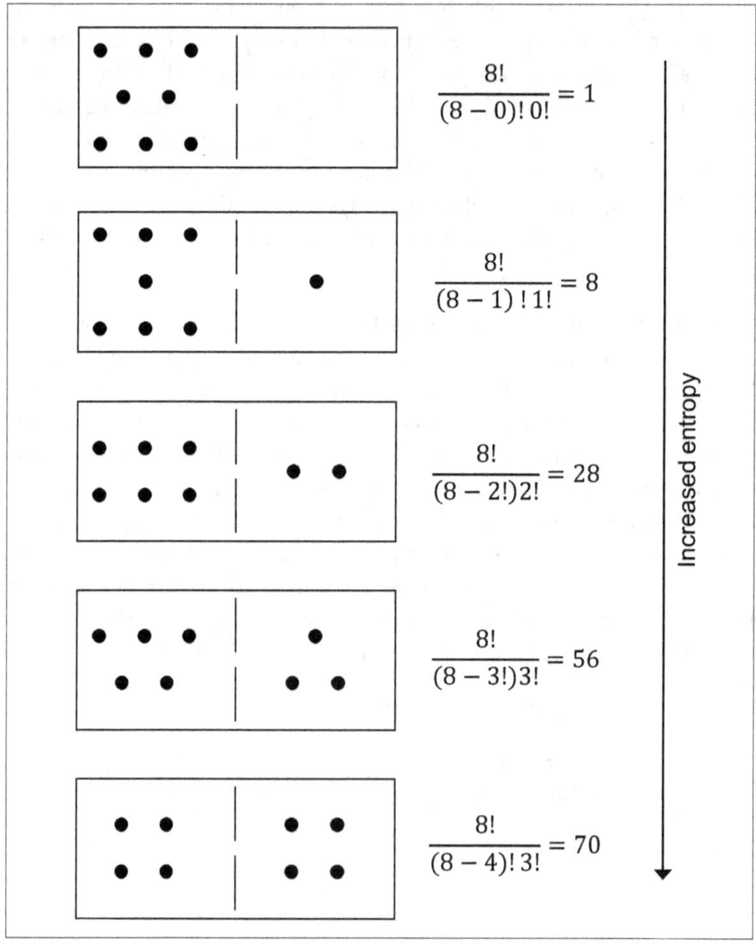

Image showing the eight particles distributed in the cardboard by the filter in the middle.

Should one particle make it to the other compartment, there are all but eight possible combinations of that happening. There is a one in eight possibility in knowing a particle position. For two particles in the other compartment, 28 combinations. For three particles, 56 combinations, and for four particles; amounting to four in each compartment, 70 combinations. This is the maximum number of combinations and, hence, the maximum state of disorder. Why disorder? Because they moved from a region where they were more concentrated to occupy other spaces. As a result, the chance of knowing which particle is in the eight possible combinations (1/8) is much easier than in the 70 possible combinations (1/70). With an increasing state of disorder, there is dwindling knowledge. This is a natural state that the particles drift into, a state of disorder. This forms the statistical interpretation of entropy, the second law of thermodynamics – where systems tend to move from order to disorder.

Now, when we merge the tendency of particles to seek mergers as low probability gambles and the second law of thermodynamics, it checks out that their actions result in the generation of entropy. Science merely describes how it happens. When Rudolf Clausius coined the second law of thermodynamics, he asserted that the natural behaviour is for particles to progress from ordered into disordered states.[13] It turns out that the fractal *behaviour* of nature extended down to small scales giving OS a lot of explanatory power. Organismal Selection does not just acknowledge that entropy happens, it explains *why* it happens. Furthermore, it explains why it is a natural process, in a slightly different way from how Clausius explained it. As I said, the theory was baffling but, surprisingly, it explains a lot of phenomena.

I expect rebuttals since I could not agree with these ideas I initially expressed. It took time grappling with them. I shall, therefore, refer the reader to what Theodore Roszark mentioned in his book, *The Cult of Information* – information does not create ideas but ideas create information.[14] What my theory explains is information from a fresh perspective. Diffusion, osmosis, kinetic theory of matter and entropy are processes, emergent from the particle's goal of avoiding annihilation. All these are natural processes. There is more to this. While particulate, small-sized organisms were trying to avoid annihilation, they created another form of annihilation for large-sized organisms. Particles struggle to exist by escaping cessation of existence. Living organisms struggle to exist by avoiding a state of thermodynamic equilibrium (i.e. death). Thus, chronologically, annihilation was initially cessation of existence that later, following the scaling up of organisms, resulted in the tumbling of organisms into thermodynamic equilibrium, a state of maximum entropy. While maximum entropy is the cessation of existence for large organisms, it is not the cessation of existence of particles. Hence, the need to view annihilation in these two perspectives: cessation of existence and in a state of thermodynamic equilibrium.

From Traits to Length of Existence

Defining evolution from the perspective of annihilation shifts the focus from what traits an organism has that are better than others, to how long can an organism survive in its universe. For a long time, the focus has been and still is on traits as in the case of genes and what they express as a manifestation of their potential and fitness. This concept is in line with the ideas and insights that drove the industrial and scientific revolutions, which focused on particular attributes of technological and scientific advancement. Is the car better than the horse? What attributes does the laptop have which makes it better than the desktop? We have moved through various types of iPhones and from Windows XP to Windows 10, from Samsung Galaxy to …you get the drift. It would, therefore, make sense that the same is seen in biology, where numerous evolutionary traits developed and preserved over billions of years are worshiped much more than technological advances.

Such a hallowed focus would explain the mystery surrounding the various features that puzzle scientists. Diamond notes that scientists still argue to date as to why male human beings have a large-sized phallus.[15] Traits also explain why scientists classify some segments of the DNA as introns and others as exons, simply by looking at the expressivity of a gene. If a segment of the DNA does not express a trait, it would be called an intron and if it does, it would be called an exon. What is surprising is that in mammals, up to 95% of their genome comprises introns.[16] But the drive for seeking out traits has led geneticists to search for the roles of introns. Increasing evidence purports that introns do serve some regulatory role and that their long lengths increase the potential for exons recombinations.[17] What OS does is shift the focus to a single question: how long can you avoid annihilation? It does not matter so much about what you have, how you are built or who your mother was, even though those traits do have a role in how long you stand a chance of escaping death. The question is simple and time-bound.

Take the example of inert gases. Helium, Neon, Argon, Radon, Xenon and all the other elements that fall in this class have the property of hardly ever reacting with other elements. A scope that only looks at the traits would only capture the atomic mass, radioactive property and other 'relevant' quantities and stop there. These quantities have nevertheless been relevant, in spite of the quotation marks. Neon, for example, has a role in lighting and Helium in balloons, big or small and in conductivity. Focusing on traits only takes it so far besides object or particle use. From the perspective of OS, inert gases have been able to attain an ability that allows them escape so many degrees of annihilation in multitudes of conditions. They are relatively stable thereby having a low probability of annihilation compared to other elements. Oxygen, for instance, can get ionised very easily. This means that oxygen has a comparably higher probability of annihilation compared to any of the inert gases. Oxygen needs to seek alternative options. It would appear more 'active' than the 'inactive,' inert gases. But this is so only because the universal pressure is much more demanding for oxygen than inert gases to avoid annihilation. Since inert gases can survive in a multitude of harsh conditions, they very rightly deserve to be called noble gases. Nevertheless, some of these noble gases still fuse to form larger organisms, such as helium that fuses in stars to form beryllium, a highly unstable element. Continuous fusion, however, results in the formation of carbon.[18] This process is known as the triple alpha process, where the helium nuclei have obtained the 'unique' name alpha. The formation of carbon, an essential element in 'living' organisms, would not have happened had there not been simple concepts such as atomic, or in this case, nucleic symbiogenesis, and as a consequence, an increase in size.

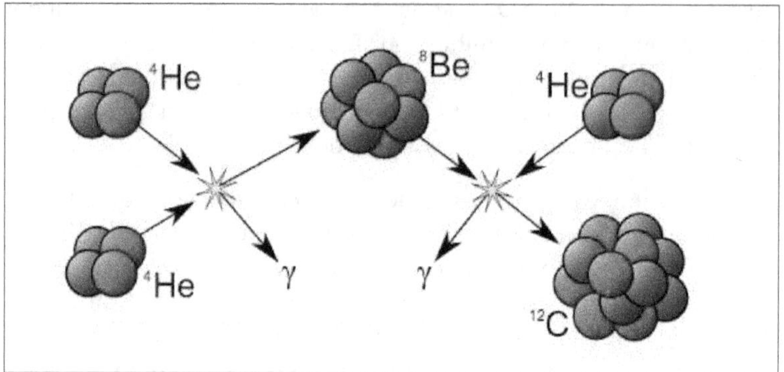

The triple alpha process. In this process, the helium, through nuclear symbiogenesis, merges to form the carbon particle, considered essential to what we understand as life.

Size and Scale

Size then became one of the single most predictive factors for organisms that I could identify at the time. While my scope was only limited to some previously unexplored consequences of relativity, Geoffrey West introduced me to other unseen size-related aspects. In his extraordinarily written book, *Scale*, I found that he had already conducted the much-needed predictive tests to conclude the relevance of size.[19] As a physicist, he tasked himself with the idea of developing a quantitative theory that would explain the features of any organism. His interests, just like that of the affiliates of the Santa Fe Institute, extended past his area of expertise. Having studied physics, a dominantly quantitative subject, he felt different and expressed his difficulties when he stepped into the biological arena, a qualitatively descriptive field. With the help of colleagues, friends and serendipitous acquaintances, he was able to develop a fairly accurate analysis of the effects of size on the organism. A much more elaborate description is given in his book. Here, I will try highlight some of his ideas in a few paragraphs.

The first thing that I got to understand clearly from West's synthesis was Kleiber's law. In the early 1930s, Max Kleiber discovered that in most animals, the metabolic energy scaled to ¾ power of the mass of the animal.[20] Succinctly, it can be stated that, if **R** is the metabolic rate and **M** is the mass of the animal, then:

$$\mathbf{R} \sim \mathbf{M}^{3/4}.$$

If the mass of an organism doubles up, that is, 100% increase in mass, the metabolic energy required would not double up but would only require an approximately 75% increase. What this means is that for every increase in size there is a reduced cost in energy consumed. There is a form of economies of scale involved. For a smaller organism there is a lot of energy required for the size compared to a larger organism. This type of relationship is called sublinear. It contrasts with a liner relationship, where there is a proportionate change in ascending or descending scale. An example of a liner relationship is that of a thermometer and increasing temperature. For every increase in temperature, there is a proportionate increase in a ratio of 1:1, a linear scaling with an exponent of 1. If a curve were to be drawn in a graph it would be a straight line through the origin (0, 0). However, sublinear relationships are non-linear. An increase in one variable does not match the other and this varies on different scales. Consider that a mouse is roughly 30g and an adult cat, roughly 3kg (3000g). The cat is 100 times as heavy as the mouse. According to Kleiber's law, the metabolic rate would scale with an exponent of ¾. That is $100^{3/4}$ making it 32. Henceforth, the metabolic energy consumed by the cat is 32 times that of the mouse, even though it is 100 times heavier. This is an exceptional demonstration of economies of scale.

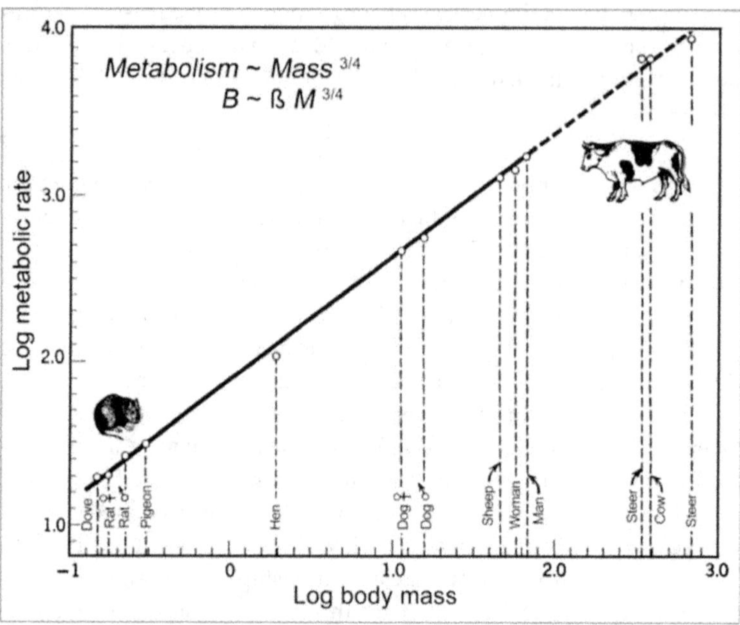

Logarithmic curve showing the relationship between mass and metabolic rate. A straight line is a typical feature seen in variables that display a power law relationship specifically in a log-log plot.

Since Kleiber's law is logarithmic, it can extend to other organisms of various scales. A cow is 100 times heavier than a cat. It should, therefore, have a metabolic energy requirement that is 32 times that required by an adult cat. The same principle applies for the blue whale, which is 100 times heavier than the cow. What of comparison between the cow and the mouse? The cow is 10,000 times as heavy as the mouse but consumes 1,000 times the metabolic energy. Compared to the blue whale, that is 1,000,000 as heavy as the mouse, it only consumes 31,622 times the metabolic energy consumed by the mouse. Since mass can easily be equated with size, there is a plausible incentive to increase in size. The Kleiber's law is straightforward – with massive economies of scale, there is incentive for increasing in size by the sublinear exponent of 0.75. This has been verified and found to be roughly accurate by various scholars and validated by West and his team.

156

This team did not stop there. While I only had an idea with limited access to data, West had both the idea and access to data. He looked at the various systems in the body and figured that there is a unifying factor of ¼, and its multiples were good approximations of various parameters of organisms. The growth rate scaled with an exponent of ¾, mitochondrial densities at ¼, the lengths of the aorta and the genome with ¼, the height of trees with ¼, the cross-sectional areas of both aortas and tree trunks at ¾, the brain sizes at ¾, the cerebral white and grey matter at 5/4, the heart rates at - ¼ and the life spans with an exponent of ¼. The negative exponent for the heart rates means that the heart rates decrease with an increase in size. As a student of the human body and its disease states, I can point out that such a relationship is a clear marker of increased efficiency. I, however, have to add that the various formulae for prescribing doses for patients does not reflect sublinear scaling. This would mean that doctors obliviously prescribe drug amounts that are unnecessary to patients or that the extra amounts could have a protective redundancy role. Studies to this effect are needed to resolve this ambiguity. Nevertheless, organisms have every incentive for efficiency and in relation to energy, efficiency increases with size. The efficiency of medication metabolism and excretion could, therefore, be adherent to this law.

What does this mean? It means that the shrew, the smallest mammal is just a scaled -down version of the blue whale, the largest mammal. Using the ¼ factor exponents, one can estimate significant features about them. It further explains why the shrew would live for only a few days while the blue whale lasts for years. The shrew is not as efficient in energy utility compared to the blue whale. The force the shrew exposes its body to subjects its cells to much more energy which not only sustains but also destroys and stresses various organs. Size, then, in a more accurate way than my version from relativity, influences lifespans and aging. The increase in size, for example, has been shown to delay aging by making the delivery of nutrients and elimination of waste efficient, thus reducing the amount of damage of the terminal units (regions of exchange). Since the terminal units are the most affected through the process of wear and tear (chemical and physical), they explain not only the uniform aging process but also why larger organisms seem to age slowly. I have a mind to think that incorporating the formulae for relativity and these exponentials discovered by West and other scientists would yield a more accurate estimate than any of them individually. This, of course, is just speculative as I do not know how to incorporate the two theorems. I only intuit that they serve a role. It could however serve as a good experiment, in the interest of seeing just how the ¼ factor laws and relativity are related.

Space-filling, Terminal units and Optimisation

In digging deeper to understand the ¼ exponents, West discovered three central aspects that I may have hinted at with regard to size. These are: space-filling, invariance of terminal units, and optimisation. Space-filling is related to a tendency to increase in size, invariance of terminal units are the unchanging interfaces of exchange of materials and optimisation is the continuous efficiency of structures. Consider optimisation, for instance. It is evidenced in so many actions done by organisms. We have already discussed how slime moulds create stalks and aggregate to form networks that allow them to scavenge for food in its surrounding. Diffusion is optimised for specific functions in differently scaled organisms. Raccoons become better at picking locks, bacteria become much better at dealing with antibiotics and humans continuously create different types of technology to solve their problems. Size in itself is the optimisation of energy manifest since only an organism with optimised growth can sustain itself.

Tied to optimisation is the idea of terminal units, which are areas where the exchange of materials happens. Examples of terminal units are terminal bronchioles in the respiratory system or capillaries in the circulatory system. From OS we already know that every particle has its unique marker of variability. However, a coarse-grained observation of the terminal units reveals their invariance across different scales of biological species in different kingdoms. Terminal units are the supply centres for a limited number of cells, hence, their invariance. The only thing that has been noted to change is that for different sizes there are different numbers of these units. Strangely enough, the numbers of terminal units have scaled with the size of the organism in the same sublinear way confirming economies of scale at this hierarchical level. It is as if there are a few infrastructural requirements in an organism's structure as it increases in size. It further confirms that the shrew, the human being and the whale are surprisingly coarse-grained versions of each other. Having already introduced optimisation and terminal units, it was the third component, space-filling, that I considered thrilling since it tied the other two, and more so because it was related to size. But first, a slight detour regarding terminal units.

Anatomically and physiologically the terminal units are universally recognized as invariant. These units are recognised due to their affiliation with various simple processes such as diffusion. However, I consider the small particulate matter the invariable units that have managed to remain largely unchanged despite the billions of years since the Big Bang. What I consider an unmatched evolutionary achievement is the ability to obtain an invariance that is so universal that it is blatantly ignored. These particles are even more stable than the DNA or its sub-units. It is almost the same as language, which is taken up so easily by creatures as ordinary that its evolutionary stability is all but ignored until linguistics emerged. Hence, my claim is that an evolutionary milestone of immense proportion is etched in the fabric of space and time if it gets to be considered ordinary or, more so, fundamental. The electron can then be considered a fundamental particle due to its invariantly changing parameters such as mass and spin. That is an extraordinary evolutionary leap! This is what I consider the invariant terminal unit – the fundamental particles at the atomic and sub-atomic scales and even at cosmic scales such as black holes and galaxies. Relative to the anatomical and physiological terminal units, these fundamental particles are more invariant than the biological ones.

Invariant is a term we use out of convenience, to make a point. What has been studied and considered invariant is only so with respect to a subset of parameters. What OS insists on is that every particle is nevertheless unique – they might be similar but they are unique. The accuracy of measurement or these parameters is currently several orders of magnitude, accounting for its accuracy. However, the evolution of measurement has shown how each theoretical measurement is an approximation of the previous one. Einstein's theory and measurement was found to be twice as accurate as Newton's theory. So, we are not certain that the weight of the electron is 1/1840 the mass of the proton but we are reasonably certain that it is close to this. It is, therefore, reasonable to conclude that electrons are fundamental particles, similar through this level of accuracy. The key word is similar. Organismal Selection agrees that they are similar but they are not the same. Each particle is unique just by mere existence.

As particulate organisms scale up, we encounter the effects of size, from a molecule to a cell, to an amoeba, a penguin, a rhinoceros and even a dinosaur. At each of these scales the merging of the lower and upper hierarchies of particles creates variably autonomous entities in the most optimised way, first by creating a dynamic form of stability as in the case of different disequilibrium states through reinforcing and balancing loops, and secondly, by making the most out of size. Size, thus, unites the terminal unit and optimisation. And the most effective way to optimise for increasing size is through fractal geometry.

Fractals and Space-filling

We have previously encountered fractal *geometry* of nature in mountains, clouds and even lung tissue. It is the ruggedness of objects that is created through a pattern of self-similarity at descending scales of an object. I have also tried to argue that self-similarity is not just seen in the shape, but also in particle behaviour; what I call the fractal *behaviour* of nature. The jaguar, myself, and the quark would have instances where they would risk their near-certain doom to clinch the low probability but worthwhile gain by seeking a merger with another particle. This is even seen among the cannibalistic praying mantis. The male is aware that they might get eaten up by the female but decides to risk it anyway.

They are almost certainly eaten for good reasons that Jared Diamond lucidly gives in his book *Why is Sex Fun*. I give what I consider alternative reasons in the next chapter but what I would want to stress is the fractal pattern of organisms, *geometrically* and *behaviourally*. With regard to size, it is largely through fractal geometry.

For an organism to optimally increase in size it has to factor in the ability to interact with as many organisms as sizably possible. Take the evolution of the prokaryote as an example. Initially, the degree with which the cell grows allows it to comfortably accommodate the basic processes of diffusion but the incentive to increase in size makes it take a turn for fractal branching. The cell wall begins to stretch and increase its surface area, which simultaneously enhances the efficiency of the cell's diffusion processes. Eventually, the increase in size creates infolds and small bits of the cell wall and cell membrane get variably exposed to the inner cellular environment and its external environment. This process would also happen inside a cell during processes such as autophagy (self-feeding) or when forming other membrane-bound organelles. The same process could help with cellular feeding other food products as seen in some white blood cells and in unicellular organisms such as amoeba. The same property could also be exploited for locomotion by creating undulating cellular membranes. This is a self-similar behaviour generating a self-similar architecture. From a membrane-bound cell increasing in size, we get membrane-bound organelles. The incentive for increasing in size would have to be matched with increased self-organization. The preservation of the basic processes would then explain the invariable terminal units such as capillaries (or alternatively, arterioles, which precede capillaries) and bronchioles in multi-cellular organisms. In all the creatures of the five kingdoms (or three, according to other schools), this property is evident – as organisms increase in size, the increase leads to an increase in the surface area for interacting with as many particles as sizably possible. These gradual changes capture the evolution of prokaryotes into eukaryotes through serial endosymbiotic steps.[21]

The self-similar architecture further captures an efficient way of occupying space, creating a fractal geometry generated by a fractal behaviour. This is the space-filling property, which can explain how the surface area of the lung is similar to a tennis court but somehow occupies the tiny space in each of our thoracic cavities. Why would an organism do that? It is simple. Firstly, by merging with another organism it postpones its annihilation through particulate symbiogenesis. The large surface area increases this possibility even though it is perceived to be in a small space. Secondly, the lung tissue largely comprises terminal units where the merging (exchange) happens. The idea of sustained merging through the exchange of materials preserves the organization of the higher hierarchical organism in spite of the particulate exchange every time an organism respires. Thirdly, the most optimal way to occupy space is through a fractal pattern. According to Charles Munger, you get a lollapalooza effect. We then arrive at the three basic properties that West acknowledged in organisms: optimisation, invariable terminal units and space-filling. According to our theory, it merely restates the connection between symbiogenesis, self-organization and size respectively. To get a better grasp of the role of fractal geometry in space-filling, let us take a deeper dive.

In mammals, the digestive tract has a series of branches that increase the surface area for interaction with small particles. Most of the absorption in the small intestine is facilitated by small, finger-like projections into the lumen known as villi. Similar projections are seen in the large intestines, for further absorption. The immensely large surface areas provided by these structures facilitate sustained merging with particles. In the kidney, there are smaller tubules that have even smaller versions of the villi known as the microvilli that serve the same purpose in absorption through increasing the surface area. The same principle can be used in the respiratory system and even the cardiovascular system. In bacteria it allows the process of diffusion to happen across the wide cell-wall surface area. In eukaryotes, such as the amoeba, it facilitates the process of food-engulfing and enhances locomotion through projections known as pseudopodia. In ants, bees and termites, it allows for maximum interaction as the foragers constitute the highly extensive interface that these super-organisms use to interact with their respective environments. To my knowledge, the maximal optimisation of space, while preserving these processes and developing new ones, happens in a fractal pattern. The cross-sectional area of the aorta as it leaves the heart is almost equal to the sum of the cross-sectional areas of its downstream terminal branches. Each branch can be quantified as a reduced-scale version of the aorta. It is the self-similar property seen in numerous duct systems. This fractal pattern, itself a feat of self-organization, would end up creating an interface for the maximal exchange of material, at the invariable terminal units. It is the hallmark of optimisation.

The idea behind fractal patterns came to the limelight following the work of Benoit Mandelbrot. In his seminal paper, 'How long is the Coast of Britain?', Mandelbrot caught the attention of the world by showing how self-similarity makes the coast of the Britain longer with every enhancement of measurement process.[22] It is this same principle of self-similarity that is taken up by biological organisms to occupy space and create the rugged picture of various organs. The brain has dips and hills that a fine ruler cannot accurately measure just like fingers have grooves and curves that a tape measure cannot capture accurately. These measurements will only be close approximations. The idea of self-similarity was then stretched by mathematicians to reflect the conversion of an object from one dimension to another. The brain, with its dips and hills can hypothetically be stretched to have an idea of how extensive it is. In so doing, the three-dimensional organ would attain a two-dimensional interpretation.

A line can be bent so that it appears as if it is a three-dimensional structure. Take, for instance, a straight wire, which is similar to a line and, therefore, only has a single dimension (1D). When it is bent so that it has four edges and four vertices such as a square or a rectangle, it acquires another dimension, resulting in two dimensions (2D). If we continue to fold the wire some more so that it has more than four edges and more vertices, it then appears to occupy space while simultaneously approaching a three-dimensional object (3D). This ruggedness of an object between one absolute dimension and the other is what is considered the fractal dimension. It is not an absolute dimension, such 2D or 3D, but a fraction between the two, accounting for the ruggedness. This ruggedness appears to be repeated variably in a similar pattern at ever-decreasing scales. It is a pattern of self-similarity at smaller and smaller scales so that the more rugged the lines are, the more it approaches the absolute dimension. This explains why the bed sheets can be compressed into tiny spaces in washing machines. A fractal dimension between 2D (when the sheet is spread out on a flat surface) and 3D (when the sheet has been folded to *almost* take the shape of a cylinder) makes this possible. Fractal dimension mathematics explains the ruggedness of reality.

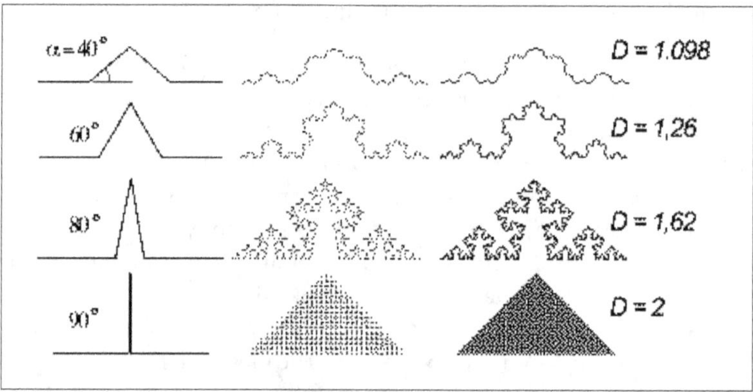

Image showing the fractal dimensions between 1 and 2. Any dimension between 1 and 2 has ruggedness as a feature. A smooth outline is seen in the second dimension, at the bottom (D=2). However, there are hardly smooth features in reality.

It might appear that the sheet is small due to the folding but when spread out, it occupies a large surface area. This is how objects occupy space in a fractal pattern. The same principle explains why the small intestine can be up to twelve meters long in a five-foot-tall lady. If you would lay out the arteries, arterioles, capillaries, venules and veins into a single length, it would be approximately 100,000 kilometres, with the capillaries making approximately 80% of this length. Consider that the circumference of the earth is approximately 40,000km. This would mean that the length of the circulatory system could wrap the earth 2.5 times! The enormous length of the genome is also able to fold and fit inside not just the cell but the nucleus of the cell, if not in the mitochondrion. This shocking feat is possible through fractal geometry. Here's the catch – living organisms display it as well as inanimate structures. The Grand Canyon and the coasts of all sea and ocean boarding countries have a fractal pattern. In fact, when one factors in the coral reefs that surround continental coasts, it bears resemblance to capillaries or even the terminal bronchioles as sites of exchange of tiny materials across a surface with an immense surface area. Clouds, mountains, rivers all display such a pattern, even though they are inanimate.

Why not call them organisms? You may argue that correlation should not be our fall-back reason to call both of them organisms. However, the various arguments I have raised thus far, including the fractal geometry of nature, should not be excluding but cumulative evidence that even what we call inanimate, have properties similar to the animate. Fractal *geometry* of structures and fractal *behaviour* of these structures are evident in both the animate and inanimate. What are the grounds for dismissal? Disequilibrium states? I do not find that argument satisfactory. Particles, the spirochete, the novel corona virus, Ronaldinho, Michael Jordan, the earth and even stars, can be considered organisms. We can even assert that the *fractal behaviour* of nature could have a hand in the *fractal geometry* of nature. Even though this could be speculative, it does not hurt to seriously consider the possibility.

Size and Paradoxes

I shall now take you back to my initial encounter with the possibilities of size playing a role in reducing or increasing the pace of life. Following the principle of equivalence, the larger organisms last longer. It reminded me of the paradox I encountered while reading Andreas Wagner's book, *Arrival of the Fittest*. In this book Wagner introduces an interesting paradox developed by the Nobel Laureate, Manfred Eigen, about the paradox of the genome.

The genome is archived in the DNA. To maintain the integrity of this structure, you would need enzymes that correct the various errors that arise during the process of DNA replication.[23] However, the structures required to do this correction are also complex structures that need development. Therein lies the paradox – to form the corrective enzymes there needs to be a complex structure such as the DNA; but to form the DNA there needs to be corrective enzymes present. Eigen managed to arrive at a threshold that has been a puzzle for biologists for a long time.[24] There too lies another problem. Eigen was a believer in interdisciplinary training and even got to learn from Nobel-winning physicists, even though he specialised in chemistry. I believe that the sole specialization could have been one of the problems that underlie the difficulty in cracking Eigen's paradox. Numerous biologists would have a narrow scope in figuring out how such a complex molecule was possible. Nevertheless, extrapolating explanations from various disciplines, OS shows that it is.

When complex molecules increase in size, they increase their chance of lasting longer. We have seen that this is possible through relativity and increased economies of scale. Therefore, the threshold that Eigen formulated contrasts with reality. It contrasts with you, an existing complex entity. I believe that the benefits of increasing size through mergers could have been the much-needed edge that organisms required to allow large complex molecules to form. These are molecules as large and complex as DNA and its corrective enzymes. The paradox further highlights the possibilities of considering simpler molecules as organisms, since it would demand such molecules and the DNA exist as separate organisms. This type of thinking might dissolve Eigen's paradox. How?

Firstly, the role of the molecule(s) would serve the DNA. Usually, DNA undergoes spontaneous mutations. Past a certain size threshold, mutations can be disadvantageous. Stability would be essential to surmount this threshold. Enter the other organism, the molecule. This molecule need not be organic. At this threshold level, the molecule would merge with the DNA and stabilise it. That is, the molecule would help minimise the spontaneous mutation of the DNA. A stable DNA would then increase the chances of elongation of the structure while minimizing the spontaneous mutation rate of the molecule. The longer the DNA, the more such molecules can merge with it. Thus, the DNA ends up serving the molecule or other similar structured molecules. Subsequent elongation of the DNA would later get to the other minimum threshold level for generating complex structures such as enzymes that would aid in DNA replication. As more DNA structures get replicated with higher fidelity, the chance of merging with the initial molecule or molecules with similar roles increases. This increases the number of options for the molecule in creating mergers. Eventually, the role of the DNA ends up later serving the molecule(s) even more. These mergers would then set the condition for creating more complex structures such as proteins. The downstream result would be the formation of larger structures such as the cell, the multicellular organism, the shrew and the blue whale.

A simpler solution would have been a merger between the DNA molecule and an enzyme-like molecule. Enzymes are organic catalysts. However, catalysts need not be organic. Therefore, the very first enzyme-like molecule need not necessarily be organic. It just needs to serve the role of stabilizing the DNA from error-prone mutations. In the event that a merger is formulated between these two entities, the error threshold generated by spontaneous mutation can be surmounted. Complex molecules can then be formed. Such a merger could have been the precursor of processes such as DNA methylation and histone acetylation. These are mergers seen in the DNA molecule and considered pivotal in studying epigenetics. Later on, the DNA could then evolve to a molecule large and complex enough to generate enzymes that would have served it in maintaining its stability during the replication process and also in the generation of enzymes.

Size brings about efficiency in energy utility. Molecules, as organisms, could have benefited from this unforeseen outcome. The large molecular structures needed the benefits that arise from size for them to form the replicative material that enabled living organisms to form. While Eigen's paradox emphasized the limiting role of mutation past a certain size, introducing mergers as a mechanism of developing diversity makes the solution of the paradox simpler than before. Our current understanding of epigenetics shows that mergers among molecular organisms (as OS describes them) enhance complexity up to and past the minimum threshold needed for the emergence of 'living' beings as we understand them.

It would mean that by insisting on the term 'living', we almost yield to some boundary threshold that blinds us to the possibilities of understanding the founding processes of 'living' organisms. Deciding to call a system 'living' and another as 'non-living' is a matter of scientific taste. This is a perspective that has grown so conventional that anything to the contrary would strike the typical scientist as absurd. However, we do not believe, but have *reason to believe* that it is a fact – that living organisms are different from other material. Unconventionally, I have reason to believe that considering other material as organisms is a viable course to pursue. I also have reason to believe that discrete boundaries, such as living and non-living entities, are almost non-existent in nature.

Arbitrary boundaries can be established in the interest of clarity. For instance, the idea of biological species differs from phylogenetic species but each has its merits in describing diversity. Each introduces a form of clarity and hence their usefulness in ecology. Organismal Selection might not have such benefits straight off the bat. Maybe later, when it has been well dissected by other scientists. Boundaries demand that we consider organisms as bounded systems, implying that they might be closed. However, they are not strictly closed but open systems. It is the only way they can interact with their respective universes. Since particles are also open systems, we can also consider dissolving the boundary problem of 'living' and 'non-living'.

Now, back to the topic of the two large complex molecules – the DNA and the repairing enzymes. To consider these two entities as organisms would demand that each has cognitive features such as sensitivity to their unfolding demise in their respective universes. They would then seek ways of avoiding annihilation through mergers. Merging, by now, should be appreciated as a cosmological strategy. The DNA and repairing enzymes have to be autonomous organisms that aligned their goals, cooperated, and ended up forming an emergent organism after the merger. Organismal Selection does not just offer a possible solution to the Eigen paradox but also supports the idea that 'biased' molecules and chemical processes should have existed prior to the DNA and RNA world. There should have been some 'bias' expressed by these particles that allowed them to form mergers and sustain their existence. The same 'bias' is enough to explain the idea of autopoiesis, which was developed by Francisco Varela and Humberto Maturana.[25] Recall that the localised *auto*catalytic phenomena and the widespread inhibitory phenomena enabled us form the robust map of an organism and its universe. The evidence and arguments Varela and Maturana put up should then support our map.

From the name, *auto*poiesis highlights the *auto*nomy of a living system. Autopoiesis refers to the production of oneself. To illustrate their idea, Varela, Maturana and Ricardo Uribe created a computer simulation of a substrate, a link and a catalyst, which they considered as the basic necessities for autopoiesis.[26] These were the rules of the simulation:

1. Two substrates under the influence of the catalyst may coalesce to form a link;
2. Several links may 'bond' to form a chain;
3. These chains can then disintegrate or merge again based on a probability factor.

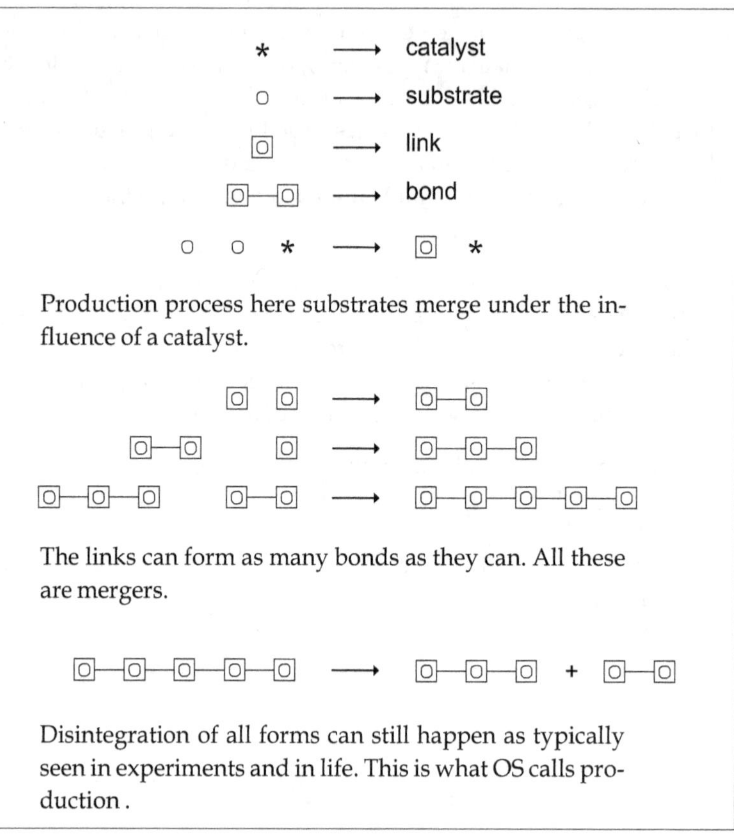

Image showing the relationships between substrates, chains and links.

Running such a simulation showed how it was able to form a closed
chain, which is similar to the membranes that bind living creatures.
Notice that the algorithms used in the simulation were in line with the
fourfold pattern of preferences, where organisms sought mergers. The
mergers used here are links and chains and interaction between a substrate and an enzyme (catalyst). While they incorporated a disintegration factor to match reality, OS can contribute to such a model. That is,
the longer the chain, the larger the size. The larger the size, the greater

the edge the organism has to last long enough to undergo the formation of the membrane-bound entity.

In keeping with the ethos of science in admitting limitations to studies, the citing of the disintegration of large, bonded chains by these workers is laudable. Such an outcome is empirically possible much the same as the difficulties that systems face in trying to maintain the structural integrity due to an increase in size. What OS adds to the picture is that the large size may be the right property that allowed the formation of the first membrane-bound cell. Size comes with the added bonus of efficient use of energy and gives the organism much more time to exist to form cellular membranes. Size could just have been the reason the initial ancestor, according to Natural Selection, survived. It gave it the necessary time to come up with the idea of producing a version of itself. Such are the reasons I fell in love with this 'S'. Size integrates the previous Ss and congeals them to create a better understanding of the arrival of the fittest. And there is more.

There is another thing that the simulation by Varela and Maturana highlights – cognitive processes. The substrate, the link and the catalyst, highlight the 'biased' tendency to merge and consequent emergence of new structures and unique organizations, what we have called new organisms. This is in line with OS. It also shows that these elements contain properties similar to organisms. They display a 'biased' sense in trying to avoid annihilation. They 'seek' mergers. This is also well described by Eigen's concept of hypercycles, which maintain an auto-preserving structure, much like the Kreb's cycle of cellular metabolism.[27] Such a tendency marches our definition of an organism: it comprises an organized set of particles that tends to avoid annihilation. It is thermodynamically open, organizationally closed and autocatalytic. Anything external to it is inhibitory (the substrates might appear to sustain it but the result of the merger is greater entropy in its universe. Even more is the fact that once an organism takes in a substrate, it forms another organism, much the same way as when we take out meals. The merger created results in greater entropy in the organism's universe as no merger is perfect). Our map still stands. This biased sense then takes another twist when we consider the works of the legendary Stuart Kauffman.

Trained as a medical doctor, Stuart Kauffman was also interested in complex systems and theoretical biology. He is another affiliate of the Santa Fe Institute, just like Geoffrey West. Leading a nest of scientists brought together by the same scientific problems, Kauffman was able to develop a model capable of explaining complex phenomena, a binary network model. A binary network is a system where an entity can exist in either of two states, conventionally known as ON and OFF. Linking these entities with inputs, he discovered that certain invariable states of order were arrived at when the inputs were smaller, often centring around two inputs. More than two inputs often tipped towards chaos and less than this would hardly yield an excitable ordered system as seen in living organisms. He called this model the NK model, where N stands for nodes (the entities) and K for the inputs.[28]

Kauffman then stress-tested his model to determine if it could capture the picture seen in gene networks. Conventionally, it is believed that the genes can be dominant or recessive, expressed or not, sense or nonsense. With these binary outcomes, it served as a good test of the NK model. What he discovered was that the idea of a genome as this unilateral molecule made of genes worked in hierarchies. Each gene was directly regulated by only a few other genes and not the entire genome in the DNA compound. The binary model turned out to be very accurate in predicting the genome networks. It is somewhat similar to the idea we mentioned in the previous chapter by Lewis Wolpert of positional information. Genes have no idea of where they are but rely on their position in relation to its neighbours to effect their role. A single gene would influence only a small number neighbouring it. In short, genes work in hierarchies. Kauffman's work adds to our assertion of hierarchies as efficient means that allows systems to work. In fact, it accounts for why systems are robust, a field that Andreas Wagner has also shown works through networks.

Since hierarchies reduce the amount of information that any part of the system has to keep track of, energy can be diverted to other pressing needs. This feature facilitates the formation of networks of various degrees between and among different parts of a system. The connection between the skin cells is much stronger than between the skin cells and the haemoglobin that supplies it with oxygen. In essence the concept of mergers shows that systems networks are as fundamental as the fundamental particles. By network, I mean how particles relate to each other. The quark exists in its stable state as a 'tripod', with asymptotic mergers that create this network. Scientific analysis focuses on dissecting material and asking what it comprises, whereas systems thinking asks how materials are connected to others. Both rigorous analytical thought and critical systems thinking are needed to understand our reality in the best possible way. Organismal Selection has tried to incorporate both ways of thinking. By focusing on the substance, we identify the organism and its properties, the gene and the behaviour, the organism and the ecosystem, the quark and the galaxy. By focusing on the relationship, we discover mergers and the map of the organism and its universe. Integrating the two, we discover the unscaled diversity that every organism wields in its universe.

Unscaled diversity allows us to briefly explore another idea that is tied to the property that we have attributed to organisms – autocatalysis. Let us assume that having existed, organisms need to continue existing. This can be considered as a baseline property of catalysis. Now, what is a catalyst? It is an item that speeds up a reaction without losing its composition. An organism possesses this property of trying to maintain its composition in a universe hell-bent on destroying it. An organism should therefore have some level of catalysis, just enough to sustain itself. It would then seek out mergers, creating more organisms in the form of more and more complex entities. With more mergers, there are more compounds and each by itself is a unique marker of diversity. If each organism has the potential for catalysis, then the potential for more catalytic reactions increases with more organisms. As these entities increase in size and number, the number of reactions catalysed increases faster than the number of complex entities or organisms present. It is easy to imagine how this may come about when a particle can enhance the production of another product. For example, if **A** and **B** can spontaneously merge to form **AB**, then **AB** has the potential to not only serve as product but also as a catalyst. Granted catalytic properties, **AB** can further enhance the reaction between **A** and **B**. It is not strange for a compound to act as both product and catalyst. Enzymes are both a product of chemical reaction and serve as cellular catalysts. The RNA as well can serve as both a product and catalyst.

Now, if catalysts only enhance the speed of a reaction without losing its chemical composition, it is safe to assume that catalytic reactions are a form of a merger. These kind of mergers between two or more compounds may be extremely short-lived. However, there is the possibility of having mergers that are long-lasting. This is possible when compounds aim to enhance their continuous existence and that of its primary constituents by serving as both product and catalyst. Assume that **A** can react with **B** to form two compounds **AB** and **BA**. We can then assume that each, by existing, tends to avoid annihilation, to maintain its structure. This we have stated is a baseline requirement for catalysis. If **AB** is able to stimulate the production of **BA** and in turn **BA** able to stimulate the production of **AB**, as long as there is enough **A** and **B** molecules, such a relationship ends up being an autocatalytic set. Interpreted in terms of hierarchical relationships, the role of **AB** serves **BA** and the role of **BA** serves **AB**. What we have is a stable merger formed from a stable hierarchical relationship. If catalysis stands as a property with the potential to create mergers, then as more molecules are formed, there is the spontaneous ability to form autocatalytic sets. This is an idea that Kauffman had where he assumed that some particles could have some form of catalytic property. What OS asserts is that they *do* have this property. For catalysis to occur, a particle has to maintain its structure – this serves as the goal of every organism, which is, to avoid annihilation. I consider this process essential in forming mergers, which is the driving force for organisms to postpone their annihilation. And how they survive is through the functions of hierarchy systems – the goal of one serves the other and vice versa. An example of a chemical cycle that displays the properties we have just described is the Kreb's cycle. This cycle has its components aiding in the production and catalysis of subsequent reactions that ends up generating the cycle, hierarchically. In the next chapter, I shall elaborate much more how mergers can increase exponentially compared to the existing particles. This would amount to what Kauffman calls a 'phase transition' into a chemical set larger than its components. It would explain how an organism can increase in size.

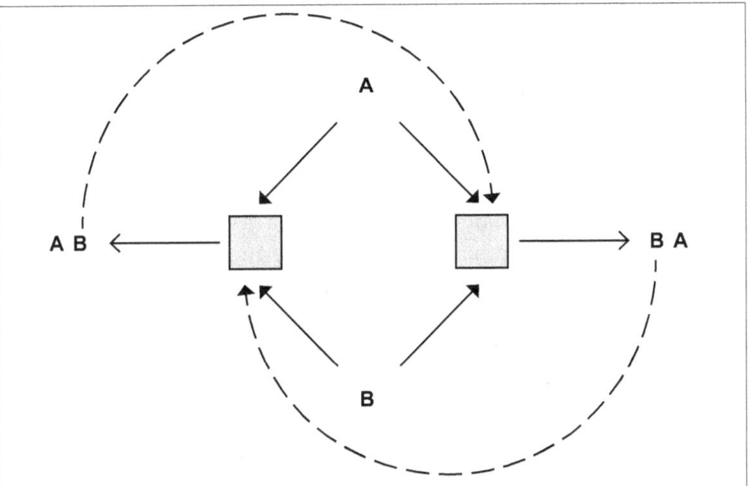

Image showing the emergence of an autocatalytic set. A and B can merge to form the product BA, which catalyzes the production of AB. A and B can also merge to form the product AB, which in turn also catalyzes BA. In the presence of a source of energy this relationship ends up being self-sustaining, a thermodynamically open but organizationally closed entity.

There is one other rather bizarre finding that Kauffman discovered from his NK model that we have only lightly mentioned. If the inputs were 'biased', there was a tendency to create ordered structures, or networks. Biased here would mean that there were nodes that were primed to be more ON than OFF or vice versa. From the definition of an organism, there is an inherent tendency to avoid annihilation. There is a bias. It could be plausible that the bias of particles to avoid annihilation is all that was needed to create the states of disequilibrium that we have come to identify as life. Adding size and catalysis to it could have given it the edge it needed just so that it can survive long enough to perform bold feats. As I have tried to explain, the bias is evident among both animate and inanimate entities. The discrete slice of calling an organism 'living' and another as 'non-living' would only make it difficult in appreciating their similarities. Even after Erwin Schrödinger's book, *What is Life?*, there was still no clear cut answer to the question that forms the title of his book.[29] It can, however, be partly solved by factoring entropy in a slightly different way. This is how.

A state of complete annihilation is a state where we cannot derive any information. We cannot derive any information from something that does not exist. On the flip side, if a particle exists there is some level of information. Cessation of existence is one form of annihilation. The other is seen in thermodynamics as a state of thermodynamic equilibrium where very little information can be derived. This is a state of maximum entropy. However, a system in a state of thermodynamic disequilibrium possesses a lot of information. Different levels of existence and disequilibria will, therefore, translate into different levels of information. Genomic studies have shown that the DNA sequence of an organism is usually unique. We, however, do not have to resort to the DNA to arrive at this conclusion. We only need to know that the organization of particles and their spatio-temporal location in the universe translates to a unique disequilibria state or existence state, that is, a unique set of information and hence a unique form of variation. It accounts for diversity.

Now, organisms can be considered to be living because they have a characteristic set of features that can be obtained within a range of disequilibria states. For instance, it is widely acknowledged that a living organism must have the potential to reproduce and self-regulate. The first one eliminates all the organisms that are infertile, inanimate particles and the Gaia. The second one includes the Gaia and the atom, the stable tripod quark, the stars, yet these dynamic states do not fit the narrow range of traits that biologists look for. As long as scientists stick to these two criteria, they will hardly appreciate the self-regulatory property of the atom or of the star; and they will not appreciate that reproduction can have different explanations besides the one deduced from Natural Selection, as I will show in the next chapter.

Since disequilibria and existence matches different forms of information, the different states of disequilibria and the existence of particles denote different orders of information. That the Gaia cannot reproduce does not mean it is not an organism. The example that I gave of the bacteria and waste products shows us that due to size, we can live without excreting products. The bacteria can live with the 'waste materials' until it has too much to handle. During breastfeeding, the newly born baby often ingests and utilizes all the milk from the adult mother without having to excrete any. The human being can contain themselves for days before relieving themselves. Urination is a slightly different thing, which we still can contain if it is below the stimulating threshold for peeing. Humans and mammals at large have developed an elaborate means that allows the rectum to contain a lot of faecal matter in this compliant pouch. When the size becomes too much, a process that does not align with our goal, it forces somebody to seek a hidden bush or lock themselves up in a room, hopefully with tissue paper.

As for reproduction, the infertile cannot reproduce but that does not make them any less living or excluded from evolutionary processes. There are many infertile organisms that have ended the lives of many fertile ones. The same infertile ones have also saved many fertile creatures from different hostile situations. As long as they exist, they have a role in evolution. I would subscribe to the notion of calling all particles organisms, and relish at the wonder that size, scaled up or down, contributes to diversity; and much more, in explaining why a child would grow into an adult and not remain a child even though they might turn out to be infertile anyway. Size would also explain why the Gaia does not appear to excrete like other organisms. In fact, size has proven to be as quantitatively as it is qualitatively predictive. West has shown how it can be quantitatively predictive but I had my own set of qualitative predictions long before I read his book. These predictions revolved around virtual particles, Gaia and cancer.

Virtual Particles, Cities, Gaia and Cancer

After I had yielded to the possibility of particles being organisms, I was still saddened by how it would sound by saying that an atom and the Earth are all organisms. How could the earth be an organism?

Luckily, Lovelock and Lynn Margulis had done their bit and found how self-regulative the earth has been since its 'inception'.[30] From their Gaia Hypothesis, the Earth could be considered an organism after all. That left me with the small particles. Seems like I was and still think that I am the only one defending them. Bearing size in mind, I reasoned that the large organisms would survive longer than the small organisms. This could be concluded from the principle of equivalence and the sublinear scaling of organisms. The reverse, I reasoned, should also be true. I then posed this assertion: there should be very many small particles all over our known universe (whatever that is) that had little time to exist and that they would annihilate very fast, some even so fast that we cannot really appreciate them directly. I was shocked, from my research, to find that this was already validated by particle physicists in what they called virtual particles.

Virtual particles cannot be perceived directly, only indirectly. That they cause a lot of perturbations was the first hint of their possible existence. They usually appear and disappear in vacuums, so that vacuums are not essentially empty as we have previously known them to be.[31] Numerous tests in quantum mechanics have been done and have validated the existence of these particles. They appear and disappear; they exist and then cease to exist. It has even been suggested that they contribute to the mass that can only match our currently expanding universe. It turns out that my qualitative prediction was right after all. If I am to validate their existence, I have to state that they should either be the same size as quarks or smaller. Their small size should explain their rapid appearance and disappearance.

To be more concrete, virtual particles appear and disappear by annihilating each other. If two were to appear, everything external to the first would be its universe, and the same would be said about the second. Since they are small, their only strategy to increase in size is through being in constant state of motion. Indeed, they are. But the universe is hell bent on annihilating organisms. Rightly so, these particles, when they collide, annihilate each other. It appears that virtual particles have high probabilities of annihilation. Consequently, they would have a relatively smaller tendencies of avoiding annihilation, in the only way they know how – motion. If these particles were the very first in existence, then a constant state of movement would be considered, according to our theory, as the very first strategy to avoid annihilation. Mergers would then emerge later and hold to be much more reliable than just continuous states of motion. However, as we have explained, organisms tend to preserve their initial properties if they serve in preventing their annihilation. The complex mammal preserved the process of diffusion at the sites of exchange as seen in terminal bronchioles and capillaries. Similarly, if being in a constant motion served the original organisms, then even after forming mergers, they should still be in a similar state of continuous motion – and they are.

As for the Earth, it is an organism appreciable by other self-regulative features and now, through size. As well, there are small particles that exist for short periods of time, virtual particles, again, due to size. My theory should further assert that Mars, Venus, Mercury and any other celestial body is an organism but not in the sense that Lovelock had identified the Gaia; only in the simple sense that they are existing and have not been annihilated. These celestial bodies have some level of organization, some form of order and, hence, substantive information can be derived from them. For the virtual particles that only make a 'flashy' appearance OS concludes that one factor that would have to explain such a behaviour is their size. Size thus has a role in organismal existence. But this is not all. There is more.

My second assertion was that the map of a localised organism and its inhibitory universe should mean that matter, when combined, would amount to a very small fraction of our 'shared' observable universe. If we are to consider celestial bodies as components of galaxies, then they are all attracted to each other by the strong forces of gravity. At such a scale, the force of gravity is immensely strong. The amount of particulate matter in the universe would then account for the universal gravitational energy. When all are combined, they would increasingly lead to unremitted attraction, exponential contraction and ultimately to one big crunch in the future.

However, the observable universe (whatever that is) is not contracting. It is expanding. There appears to be another force that is just the right amount to let the universe expand. It is similar to the critical velocity that rockets have to attain for them to escape the earth's gravitational pull. Einstein called this force a gravitationally repulsive force and added it to his equation, the famous cosmological constant. It has qualities similar to that we have already described as inhibitory; it is repulsive, while matter and, hence, gravity, is attractive. Even more is that the amount of matter, which is always attracted to each other through an inverse square rule, is a form of a merger. By this definition, all matter constitutes one organism, merged through gravitational force. Anything that is counters that results in resistance. The credible threat is repulsion, hence the counter force of attraction by the cosmological organism. All particulate matter is one single organism united by the inescapable merger forged by the force of gravity!

What of its universe? Is it as widespread and as universal as any other map we think? It turns out that the amount of matter present in the observable universe constitutes roughly 5% of the universe. This would make its universe 95% - indeed, this is widespread, while the organism can be classified as localised from these percentages. Furthermore, this universe comprises material that we do not quite understand. For this reason, it is usually regarded as dark matter. Some posit that dark matter, and dark energy, may be responsible for the gravitationally repulsive force.[31] All in all, it does turn out that my predictions were somewhat fitting. The widespread scattering of the galaxies and matter in our 90 billion light-year-wide, observable universe reminded me of the super-organisms, who had foragers and soldiers scattered around their environment despite being a single organism. These findings were so exciting that it got me thinking if I was just forcing it all to fit into a Procrustean bed. The map of an organism and its universe, respectively localised and widespread, one with a tendency to exist and another with the intention of annihilating, applied even at astronomical scales. This finding was thrilling. And there is more, but this time, in my field – medicine.

In my second year of medical school, I got entranced by a group of cells that were used in the treatment of cancers. These cells were known as hybridomas, a hybrid of immune cells and cancer cells. Their role? Detect the specific cancerous cells and eliminate them from the body.[32] During such moments I would toy with the idea that what we call cancer might be an evolutionary trait that humans cannot contend with, a step that they cannot completely do away with despite the various efforts at curing it. Nevertheless, in the spirit of avoiding human annihilation, several institutions have been put up with curative intentions. On one end, some cancers have been successfully treated; on the other, the annual budgets of countries have cumulatively contributed large sums of money towards curing this elusive malady.

Numerous studies have linked cancer to the increasing adoption of Western dietary habits such as high sugar and high fat diets.[33] Habits such as smoking have now been established as significant causes of very many types of cancer.[34] Granted, there are very many chronic smokers who do not end up getting these cancers, though statistical interpretation tips smoking as a high risk if not a significant cause of cancer. In as much as this is hardly disputable, it begged the question: why do mice have very many tumours? It is not that mice also smoked cigarettes. Tumours are swellings that can be caused by cancerous growths. Mice have been used as models for testing the effectiveness of cancer drugs and rightly so due to their high predisposition to tumours compared to humans. However, when scaled for size, it is seen that mice have more tumours than humans, and humans have more tumours than elephants or even the blue whale. In explaining cancer, I then made the other assertion, in relation to size, that organisms would tend towards an increase in size. One of the ways it does this is through cancerous growths. Therefore, the smaller animals would have more cancerous growths and the bigger ones would have fewer if not none. As innocent as my prediction was, I investigated my deductions and was dumbstruck by the results.

The conventional idea following the genetic view of cellular function was that more cells should expose animals to cancer. Since there were more cells, the reasoning was that there should be more cells available for cancer to form in larger organisms. This is not what Richard Peto, an Oxford statistician found out. He discovered that the cancer rate incidence did not appear to positively correlate with the number of cells of an organism. It was the opposite – the smaller the organism, the higher the incidence rate of cancers and vice versa. This peculiarity is known as Peto's Paradox.[35]

Different scientists have come up with ways to explain this phenomenon. The formula by Geoffrey West hints at the increased wear and tear a small organism is exposed to compared to a larger one, which has more efficient use of energy. It is likely that exposure to such energy rates can induce mutations translating to more cancer incidence rates in mice. I have a different hypothesis tied to the organismal tendency to increase in size: smaller organisms would have more cancer rate incidence than larger animals. Cancer, itself being an organism, would have a predisposition to increase in size. On a relative basis, the smaller organism would be pressured much more to grow bigger than a larger organism. This solution is not just simple but also highlights another perspective about organisms – that they are hierarchical and have some degree of autonomy.

Firstly, the morphology of cancerous cells can be identified using fractals. The fractal branching patterns of vessels in a cancerous growth are different from normal body cells. They are much more space-filling, with higher fractal dimensions than normal tissue.[36] One can then use the disparate fractal dimensions to know the demarcation of a cancerous growth from a normal healthy tissue. A higher fractal dimensionality means that the tissue occupies more space – the hallmark for an efficient increase in size. This is the first evidence of its autonomy. The other is the metabolic processes. The central one that I will highlight is the Warburg effect.[37] The Warburg effect is a metabolic process in cells that differs from the typical metabolic process a cell undergoes in the absence of oxygen. Usually, when oxygen is absent or reduced cells shift from aerobic to anaerobic forms of metabolism. Anaerobic metabolism results in the production of fewer energy-rich molecules than aerobic metabolism. The Warburg effect also uses anaerobic metabolism but produces more energy-rich molecules than a typical cell its anaerobic state. The cancer cell can therefore produce more energy molecules even at reduced oxygen levels. The unique form of metabolism also highlights the organismal self-organization capability, arguably from the same genetic make-up. Fractals and the Warburg effect highlight not just the autonomy but also demarcate the hierarchical boundaries of cancerous growths. These growths are a manifestation of an organism tending towards an increase in size, which from its high fractal pattern bears with it the benefits of scaling. The benefits are also clear from the very idea of having cells increasing in size even under reduced oxygen levels. Hence, the drive for organisms to increase in size would expose the mice to a higher incidence of cancers compared to humans, and humans much more compared to blue whales.

Why then do humans die? If its growth is marked with efficient steps at increasing in size, why then, does it end up killing people? Hierarchies. Remember for a stable hierarchical relationship, the purpose of the upper hierarchies was to serve the lower ones and the lower ones to serve the upper ones. In the case of cancer, the lower one just served itself at the expense of the upper one. Therefore, an effective means for the animal to survive is taking out the cancerous growth. This option validates the surgical methods and attempts at nanotechnology and laser surgery.[38, 39] If, however, it is impossible, then the organism, cancer, continues to serve itself at the expense of the individual patient. If the upper and the lower hierarchies do not serve each other the system is bound to collapse, ergo resulting in death. Interestingly, the death of the system usually does not dictate the death of the cancerous organism. Some cells continue to grow. Some scientists have even preserved some of these cells in the interest of understanding their mechanisms and creating medication against them.[40] This is further evidence of the autonomy of organisms at various hierarchies. Cancer, then is the sad price smaller organisms pay at the expense of trying to do it big. You literally go big or go home.

The principle used by cancerous organisms is similar to what any other organism does. They tend to increase in numbers and hence, in size, dictating the sequence of events seen in many disease states. You can acquire an infection long before it manifests symptomatically. Some people would also get the same infection but do not express the symptoms. Should we then call these people ill? Even by WHO's definition they are not. What does that say about disease states? What does that highlight about hierarchies? You are only considered to have a disease if it affects your normal wellbeing; that is, the stability status of hierarchies. There are certain mergers that would impact you but would not impact me. When a baby is born breastmilk introduces the system to a variety of microbes that create mergers with the baby, allowing it to survive for a long time and, hopefully, into adulthood. The microbes establish stable hierarchies with reciprocal purposes that serve each other. As long as the organisms have overarching goals that are significantly aligned, the hierarchy will be stable and existence would be prolonged.

This does not just stop at disease states but in all types of mergers. Let us say a man and a woman give birth to a child. In such a family, we visibly have more than three organisms: the man, the woman, the child, and the emergent organisms that form as a result of various combinations of mergers among the three. The mergers are evident from the means of communication, where the parents coordinate their actions to match that of the child. They may decide to whisper when the child is asleep or restructure their daily routines to include the newly born. The purpose of one serves another. The merger is evident and a new organism manifest. However, communication through the coordination of behaviour is what I call weak testing of the existence of an organism, even though it admits its presence or absence. The stronger test is subjecting the organism to a form of a credible threat. Try to threaten a family and sibling rivalries dissolve. Try to break a couple and counterintuitively they often end up being stronger than before. Try to break a parent from a child and you have a stream of lawsuits or a fight. But, remove the credible threat and sibling rivalry ensues. Parents quarrel. And even the mother gets exhausted by the crying baby. These states of conflict give the impression of disparate individuals but when tested they show that they are in a merger and that there are more organisms involved.

Separation should then result into a lot of pain for the organisms since this is counter to their evolutionary intentions. If the mother or the spouse abandons the child, several mergers dissolve. Since a merger is a form of self-organization, it ranks lower than the goals of organisms. Spouses or even family members would therefore have various reasons for leaving, which are tied to their individual goals. Goals are a much stronger drive for the behaviour of organisms than self-organization.

Subjecting an organism, if you suspect it exists, to a form of credible threat reveals two important things. First, it takes you from speculation to confirmation. Where you initially speculated that there might be an organism, the test confirms or denies it. Secondly, in the absence of the form of a credible threat, different organisms display their autonomous selves by occasionally engaging in conflict, which is a scaled-down version of a threat. In short, even the strong test of the presence of an organism reveals a fractal *behaviour* of organisms. Researchers have called the behaviour of families or groups uniting in the presence of imminent threat group salience, a concept I got to learn from the work of Judith Rich Harris. A caveat from this concept was that it only described groups. When scaled down, I saw that it can apply to individuals too. But individuals were organisms. And groups with overarching goals, according to OS, are organisms. Rather than call it group salience, we can call it *organismal salience*. Hence the single best test of verifying the presence of an organism is by subjecting it to a credible form of threat.

The above examples also hint at how organisms behave – they tend to increase in size. A biological organism primarily does this through physiological growth. Growth is then coupled with increasing complexity occasionally resulting in the development of more efficient processes. Growth and development would then peak in adulthood. Reproduction would then be the secondary means of increasing in size. What would then follow would be the division of labour, and a redefinition and distribution of roles. The division of labour also happens in the developing child where one form of haemoglobin changes as they grow into a more efficient one. The organs that produce blood cells also change over time. Division of roles has a fractal behaviour in the primary growth in size, physiologically, just as it is seen in the secondary growth, reproductively.

The same mechanism to increase in size reproductively underlies septicaemia, that is, when bacteria find their way into your circulation and cause sepsis. As for humans populating the earth and causing 'geological sepsis', Lovelock calls it *primataemia*. This tendency to increase in size is so intense that it resulted in the formation of cities. Cities portray a feature that epitomises this undying need to increase in size. West considers cities differently based on their behaviours but I consider them as master organisms that have developed ingenious ways of increasing in size. A little insight into the conversion of man from a biological man to a 'socioeconomic' man would elucidate man's obsession with size and thus, his association with the development of extremely robust organisms that are cities.

Before the advancement of technology, improved health care, and development of cities, the biological man would live for an average of 30 years. If you were lucky maybe 40 years. The industrial and scientific revolution would later increase man's welfare to the point where the lifespan of human beings more than doubled. This is a weighty contribution that West discovered when he plotted their line graphs and found that man was an outlier as far as he was a 'socioeconomic' being, while the other living creatures were 'biological'. He cites evidence from, among other countries, England, which has a fairly presentable and reliable record of statistics preserved for a long time.

England's life expectancy has been about 35 years from 1540 to 1840 – for over 300 years! It reached 52 by 1914, just when World War One was starting. It rose to 63 in 1940 and shot up to 81 at the turn of the century. It currently rests at 79 years for males and 83 years for females.[41] Such a shift only speaks of a human need to avoid annihilation. In the USA alone, over 50 billion USD is used on ventures towards postponing death with key contributions from magnates such as Peter Thiel and Larry Page toward not only the treatment of Cancer but also increasing humanity's lifespan – the metaphorical fountain of youth. However, there is a cap to which humans can live, and again, it is limited by average size of the human being. West and his team managed to calculate it at roughly 125 years. At 122 years, Jeanne Calment, from France, is the oldest person recorded to have ever lived, while Kane Tanaka, from Japan, remains the oldest living individual at 117 years (this might change by the time you read this book). Little has been done to identify elderly persons in other continents and whether they conform to this size-related age cap. While all this shows how predictive size can be used in capturing particulars about organisms, the question of prolonging one's stay whereby an organism extends its lifespan, is one that I am yet to find intensively tackled in texts touching on evolution.

According to the prevalent understanding of evolution, once one has attained the reproductive age and has given birth to as many offspring as possible, their evolutionary journey is moot. Such a perspective would, unfortunately, eliminate the infertile ones out of the picture long before they reach adulthood, even though they are very much involved in the evolutionary tapestry. It would also eliminate the women that have hit menopause, or the old organisms that are not capable of reproducing anymore. Yielding to existence rather than reproduction would, however, include particles and, hence, all organisms as those subject to the rules of evolution. This would then explain the relevance of having the old people in a village or any given community.

The elderly are the library of the community, which, through the mergers created by their language, convert the community into an organism. If the elderly are archives that store information for the community (organism), then they are equivalent to the genomic library archived in the DNA. By striving to stay until old age, these readily accessible members can help the community during times of famine, illnesses, or even war. They can give sound advice gathered through experience to the prevailing leaders and even take part in raising the young members of the community. Their relevance in a community comes to the fore long after their reproductive capability – they can keep the organism (community) far from annihilation. Their role in the maintenance of the organism that is the community explains, among other reasons, why many communities do not do away with them.

What of the biological species? This is an interesting one. Biological species are a group of organisms that can interbreed in natural conditions to give viable offspring. The ability to interbreed gives a biological species two things. Recall that following OS, a particular biological species is an organism in itself. Therefore, these organisms tend to increase in size largely by secondary means, which is through reproduction. Secondly, biological species create a closed gene-pool. Only the members of a particular biological species can *ideally* interbreed with each other. They are bound, a system that is organizationally closed but thermodynamically open. One major thing they have going for them, in part, is increasing in size through reproduction. Whichever means of increasing in size improves the chances of more resilience and efficiency. In doing so, it introduces a fresh perspective about species but also a reversed way of interpreting members of a food-chain.

Before we dig into the switched interpretation of how food-chains can be interpreted, some specifics about speciation are warranted in the interest of highlighting another 'switched' perspective about living organisms. The normal human being has 46 chromosomes (diploid) in all their cells except the sex cells that have half that, that is, only 23 (haploid). The interesting point to note is that in these two groups of cells, the diploid and haploid cells, there is a progressive tendency to increase in size. In females, the number of ova multiplies *in utero*, while sperms continue being produced until men die. For women, this size allows them to have a chance at making a baby by regularly releasing an ovum in a span of 21 to 35 days. In men it allows them to form mergers and, hence, babies throughout their lives. Now, definitively, species are organisms that have developed intrinsic isolating behaviours that exclude them from freely interbreeding with others. But according to OS, species are organisms, organizationally closed, and thermodynamically open. Combining the two concepts, sex cells are completely different organism from body cells. Each continuously seek strategies of increasing in size, and they do by cellular division in the gonadal organs – ovaries for females and testes for males. And as two organisms, the haploids and diploids, do they have a merger? Yes. Blood gonadal barriers. The barriers minimise immune attacks of the diploids on the haploids. Furthermore, one organism serves the purpose of the other. The increase in size of the haploids increases the chances of the diploids to survive and vice versa. This is a stable hierarchical relationship.

Even more, is that in polypoid hybrids their only solution is increasing in size by doubling their chromosomal number. Take the example of the radish and the cabbage. They can breed and create a hybrid with three sets of chromosomes (triploids). However, three cells cannot comfortably divide into the sex cells (sex cells typically have half the set number of chromosomes as the other cells of the body). It would be odd (pun intended). Their solution lies in doubling the number of chromosomes so that three sets become six, and this can then form sex cells, turning hybrid infertile organisms fertile. It makes it all even (pun intended). This example might be a coincidental one, where doubling is a viable solution. It nevertheless shows that it incorporates size, which offers solutions in different platters. And speaking of platters, before we deviate further, let us revert to food chains.

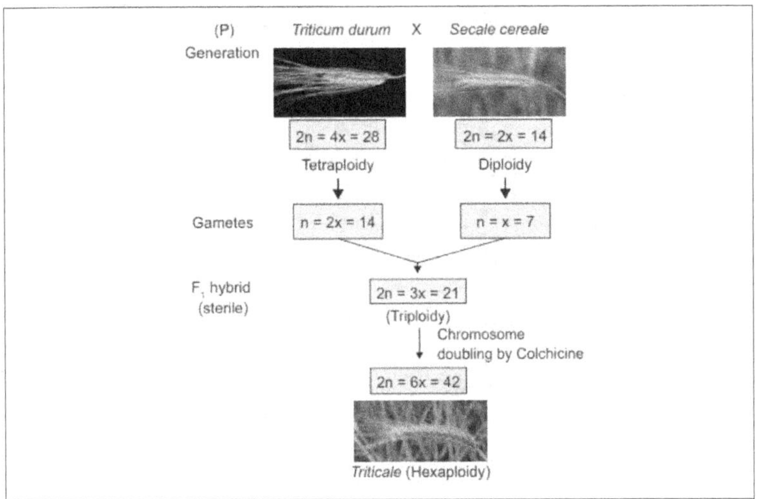

An example of how two organisms, durum wheat (*Triticum durum*) and rye (*Secale cereal*) with different genomic pairing (tetraploids and diploids) can result in a hybrid organism, (triploid) incapable of forming gonadal cells. However, by duplication (as seen in Ohno's Law) these hybrid organisms can reproduce. The cabbage and the radish are a single example among a multitude of organisms that manifest this process.

The members of a food chain can be divided into producers and consumers. The producers can be plants and consumers can be scavengers, herbivores, omnivores and carnivores. Apex predators such as sharks and lions are considered the most dangerous in the wild, but according to size, they are the most fragile. This can be seen from two lenses. The first, using the rule of 10; plants utilize 10% of the sun's energy to produce food, herbivores consume them to make 10% of energy and carnivores take 10% from the herbivores. In short, carnivores take a small portion of energy derived from the sun. Secondly, they are the rarest of the species. Tigers are mostly located in a few Asian countries and hardly on most continents. Lions are largely located in the Savannah grassland. Sharks are small in numbers compared to the schools of fish that they feed on. Considering the individual apex predators as biological species and hence as an organism, they are small. They are, therefore, less efficient and less resilient than those on the flip end of the food chain. A slight change of the environment always has them easily endangered if not extinct, as in the case of the dinosaurs, particularly the apex predators of their time. Phytoplankton and plants would then be the most resilient compared to the sharks and tigers. Humans would be the most resilient among the apes. The resilience of viruses would only rival that of bacteria. Among the apex predators those that survive in numbers, such as lions, are bigger organisms in terms of size and, thus, more resilient than the solitary tigers. Size, then is a means of postponing annihilation of an organism in the sense that OS describes it.

Incorporating size and self-organization further reveals how the goal of avoiding annihilation allows the formation of hybrids or 'semi-species'. Goals drive self-organization. Some plants, for instance, can have their pollen broadly scattered and fertilisation still takes place in another plant of a different biological species. The more stable strategy is seeking mergers, even if an organism forges it with another from a different species. In short, what size and self-organization does is that it reveals the relevance of a systems perspective. From systems thinking, we get to understand that a community that shares the same language is an organism, comprising autonomous members, who nevertheless depend on each other in one way or another. It also allows us to consider biological species, conventionally useful in assessing the biodiversity of life, as an organism. By emphasizing individual components rather than the relationships between them, analytical thinking hardly creates enough room for this possibility. Hybrids, however, exist. They are existent due to the possible formation of mergers that have been existent during the era of microbes. Lynn Margulis calls it the acquiring of genomes (top-down perspective) unlike the idea of mutation of a single gene (bottom-up perspective). Incorporating the two ways of thinking shows the relevance of both in getting a better understanding of 'living' and 'non-living' phenomena. It brings to the fore the idea of perception – how the two schools of thought perceive phenomena. This is in line with the Popperian schema on how solutions, encased as tentative theories after elimination of errors, only yield more problems, appreciated or otherwise.

Perception of phenomena extends across all the scales for the organisms. For instance, we can think that the house fly moves very fast, but the housefly likely posits that we are a strange set of beings that move slowly. In fact, we tend to think that the observations we have made are immediate even though there is bound to be a delay from the time light bounces off the object and hits our retina to the time it is being decoded by the brain. The illusion of immediacy becomes more obvious when we 'think' that our auditory and visual perceptions match. The stimulation of the visual sensation takes on a different time and path compared to the auditory one.

When we switch this up we can imagine the Gaia having a totally different experience. At its level, it 'perceives' the speed of light as normal or at times slow, when at our level we view it as the fastest thing we know. At the galaxy level, the Milky Way definitely considers the speed of light to be slower than how we perceive it. What is more is that at these levels, light is not separated from the galaxy – it forms part of it.

Similarly, the human being would not be able to exist comfortably without the normal flora on its skin, or those lining epithelia in various parts of the body. Some even produce molecules that influence perception and moods. Constant intake of some antibiotics can affect the bacteria lying in the digestive tract resulting in diseases that are easily managed by simply ceasing the intake of that particular antibiotic.42 Systems thinking shows how the human being does not exist without other bacteria or even viruses. By forming a merger, there is an increase in size, which gives it the much-needed edge to avoid annihilation. It forms hierarchies with many other organisms that when stable create a resilient system which arguably, validates the name 'human being'; the emphasis is on 'being'. On a smaller scale, this human being comprises a fairly consistent genome punctuated with variant karyotypes that aggregate and combine to form a stable system, an organism, with the upper hierarchies serving the lower ones (humans not taking antibiotics or not constantly sanitizing) and the lower ones serving the higher ones (when bacteria maintain the integrity of epithelial linings). The human being is a 'being' because of these mergers, which tie the roles of *symbiogenesis*, *self-organization*, and *size*. The right mixture explains how even *statistically* an organism is the fundamental unit of diversity. The 4 Ss are here with us again.

A little modesty is, therefore, needed. And in the spirit of modesty, OS appears too simplistic to explain the multiple variant features seen in nature. I do not deny this. Models nevertheless have to be simple for them to clarify in their simplistic way, the complex phenomena that we try to understand. In the same spirit of modesty, OS tries to picture how the subatomic particles behave similarly to biological organisms. The same applies to stars, which maintain a delicate balance between attractive force of gravity and repulsive force of nuclear fusion, another manifestation of self-organization and nucleic symbiogenesis. This is the fractal *behaviour* of nature at all known scales. Organismal Selection, therefore, takes the question about the possibility of extra-terrestrial lifeforms and redefines it to: Could there be other entities that understand the universe differently besides our way? The simple answer that OS tells us is *yes*. Our senses have for a long time believed in the deterministic model created by Newton. Yet the most tested and still robust theory of quantum mechanics feels odd. And in the same spirit of theories that appear odd, and hoping you have broadened your perspective for this next bit, I would like to go back to the one aspect that West has decided to major on later in his life – cities.

The emergence of cities all over the world reminds me of the emergence of organisms. West has been keen to notice that the biological organism differs from cities from the scaling property he used. Biological organisms increased in size in a sublinear way, so that they maximize on the economies of scale. At the exponent of 0.75, the Kleiber's law is an example of this sublinear scaling. Cities, however, scale super-linearly, at an exponent of 1.15. They are open for growth and scale not exponentially but super-exponentially.

Super-linear scaling in cities regards the *socio-economic* interactions, covering aspects such as city patents, capital generation, diseases, employment, crime rates, diseases and rates of slum development. In short, the super-linear expansion covers both the good and the bad. Therefore, while the population of a city increases, the socioeconomic aspects increased much more, at a super-exponential scale rate of 1.15. However, regarding *infrastructure*, cities scale in similar ways to an organism, sub-linearly, though at an exponent of 0.85. Thus, cities benefit from the economies of scale with reference to physical infrastructure. The more the city becomes more populated, the more efficient its service delivery becomes. It can have the same 'terminal units' serving more people compared to times when it was smaller. A petrol station that initially served several hundred city residents would then serve several thousands, rather than have the mayor put up a new one. With size, infrastructure use becomes optimised.

At this point, I could not help but think that even among biological systems, there are similar 'socio-economic' interactions that also lead to structural efficiencies. Signalling pathways, for instance, could constitute the means by which the organisms at different hierarchies interacted. I speculate in this sense, bearing in mind the idea that coordination of behaviour constitutes communication. Enzyme-substrate specificity and hypothalamus-pituitary axes with various parts of the bodies are some of the 'socioeconomic' interactions that I can only speculate about. Since West's analysis of structures of the body, such as number of arterioles per organisms, was *physical*, I figured it should be possible that for organisms – that is, organs, tissues or cells – that interact, the 'socioeconomic' interactions should also scale super-exponentially. One factor that stood out was the scaling of white and grey cerebral matter that scaled at an exponent of 1.25! Brain activity is a nexus of communication, an epitome of interaction containing different systems with the highest degree of optimisation known to man. And its networks scaled super-linearly. It is the only hint I can make that even among organisms the 'socioeconomic' interactions were present and scale super-linearly, so that the different organisms at different hierarchies could *rationally* contribute whole-heartedly. Why? Super-linear scaling results in rewards that are not zero-sum. If I dedicate 1/5 of my energy, I will not get 1/5 in return but much more. The study of cities reveals this much. It is similar to the idea I previously described, where catalytic reactions (interactions that make up the 'socio-economic' aspect of organisms and hence, mergers) supersede the formation or generation of organisms (the 'physical' aspect of organisms).

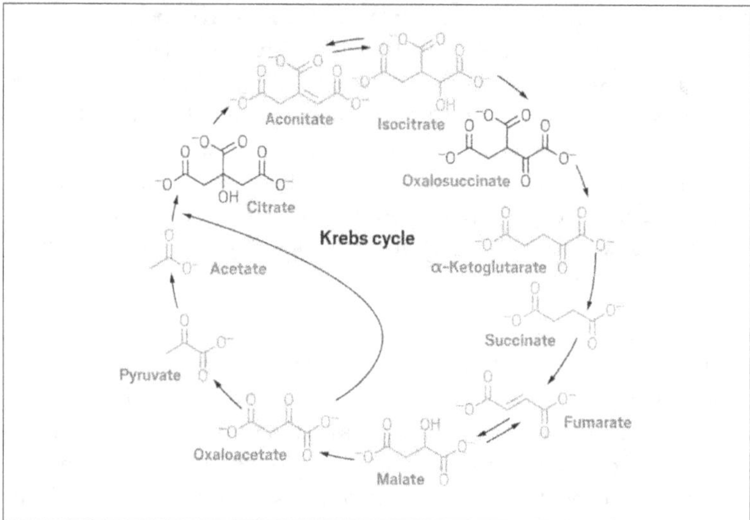

Image of the Krebs Cycle showing how catalysis and mergers can sustain systems.

In cities, there is an amplification of returns, which can perhaps contribute to the remarkable efficiency in its infrastructural use. I reckon that it explains why even creatures such as slime can come together from single cells to form a special merger that allows them to forage for food knowing that their efforts are only but a fraction of the rewards. Size, then, appears to foster much more than the economies of scale, which is appreciated *structurally*. But *behaviourally* there is likely to be communication constituting the 'social' aspect of 'socioeconomic' interactions. These interactions could then be increasing at a super-exponential rate, allowing for organisms to form all types of mergers. All in all, there is much more incentive to increase in size as little effort is matched with greater reward. From this point of view, I would still identify cities as organisms for the above-mentioned reasons and more.

I have already mentioned that one way of detecting the presence of an organism is to subject it to a credible-enough threat. Cities have been able to bounce back after so many threats. Hiroshima and Nagasaki grew into the large cities they presently are even after the bombing in World War 2. Cities in developing countries have remained resistant to the economic states of the countries. In fact, it appears that these are the very entities that often save countries.[43] However, as I mentioned, the socioeconomic aspects are not just the good. There is also the bad that stems from it such as higher cases of respiratory diseases, crime, unemployment and the rise of slums. It is perhaps another manifestation of Popperian schema, that the solutions are never sure-fit fixes. In relation to the Gaia, slums can be equated to 'waste products' of cities, but the cities cannot 'excrete' them. The size is very often manageable for the efficient cities, much the same way several 'waste products' are manageable for the Gaia. What it further stresses is that cities have their own 'metabolism' or what might be recognised as microclimates. It might be raining in Nairobi City but a thirty-minute drive from the city centre might reveal something different. The atmosphere and metabolism of a city is akin to the atmosphere and metabolism of the Gaia – they are self-regulative. The one major difference might be that cities can very visibly and ingeniously grow in size, but that might be impossible for the Gaia.

If cities are possible organisms, what of ecosystems? We have Lovelock's evidence to support the notion that ecosystems are the various organs of the Gaia. They are therefore organisms in their own right at their hierarchy level. The ecosystems are akin to organ systems. In the same way, there is a circulation of blood in the mammalian body to circulate nutrients and eliminate 'wastes', the Gaia presently contains carbon dioxide. The circulatory system operates using cycles – systemic and pulmonary circulations – just like the Gaia, which has the Carbon and Hydrogen Cycles. To quote Lovelock, 'carbon dioxide is the key metabolic gas in Gaia'. It is responsible for climate change as seen in the global warming effects, in plant growth and in the generation of oxygen, that sustains several kingdoms in the Earth's planetary system. It runs contrary to what we have always been taught in primary school: that roughly a fifth of their air comprises oxygen, the active part of air. Carbon dioxide stands at a meagre 0.03% yet it proves to be much more important than oxygen. Indeed, with reference to the life of Gaia, it has to be the most involved and, hence, the most active part of air.

For years, Lovelock noted the 'metabolism' evident in Gaia. Metabolism is a central feature that biologists consider relevant in categorizing an entity as either animate or inanimate, living or non-living. Some scientists cite this caveat to rule out viruses as living organisms.[44] If I have not stressed this point enough then I shall repeat it: Lovelock would not have been able to solidify his defence were he not able to dismiss the boundaries of academia. With the help of Lynn Margulis, who was and still is acknowledged for her contributions in microbiology, they were able to develop the Gaia hypothesis. Even though we now celebrate their victory, there is still a prevalence of the disease state that belies present fields – an opacity created by hyper specialization. It often consists of a team of professionals who seek to review work so that it fits the taste of their respective fields. This is not a bad thing. There have been numerous anecdotes of novel findings by digging deep into subjects. However, had crystallography not penetrated biology, there would have been little knowledge about the double-helix DNA.[45] Had technology not permeated biology, genomic sequencing would be difficult if not impossible. Problems that we find in the world do not give a care about the specialised boundaries we put up. They are chaotic and wild. Thus solutions, E.O. Wilson argues, can be found at the intersection of these boundaries.[46] Such are the reasons I use to consider the slim possibility that particles could just be scaled-down versions of organisms with their very own attempts at dynamic stability constituting their 'atomic metabolism'.

It should, therefore, not be the case that particles be considered dead simply because they lack the qualities we see as essential in life such as metabolism. At our human-size scale, our abilities are a smidge too limited to consider, with fair precision, what is living or non-living in the small-scale world of particles. It also breeds a lot of animosity between groups. I would advise that we briefly forget the distinction to consider the possibility. Take diffusion and osmosis, for instance. These are simple concepts that no one, to my knowledge, has seen the need to investigate further beyond the random behaviour of particle movement. If they did, it would likely have remained unappreciated or even shun. Another example is autopoiesis, which has been considered by Varela and Maturana as adequate in the description of life at its inception. However, no biologist would agree that a micelle is a living organism despite the fact that it meets all the criteria for autopoietic systems. Pier Luigi Luisi has even succeeded in formulating chemical autopoietic systems in the laboratory, but these systems can hardly be accepted by biologists or even the commoner as life bearing.[47] Since life is a difficult concept to define, I would rather that we stick to organisms as I have defined them. It allows for there to be 'life' whatever it means, and however anyone wants to define it, from entities that do not poses it.

To this point, I have spoken about size with regard to space-filling. The most efficient way of occupying space, while maximising the potential for mergers, is through fractal geometry. However, there is another bit I have to implant in you, the reader, about symbiogenesis. Growth is an increase in size, which can be considered a single scalable dimension of diversity. Development, however, is an increase in complexity, which is an unscaled dimension of diversity. With each developed merger of whichever type, there is an increase in complexity. Particulate steps of growth and development started a long time ago when hydrogen nuclei merged to form helium and several helium atoms to form carbon. These same steps explain the complex albeit increased size of the Mimi virus, whose size is comparable to that of mycoplasma bacteria.[48]

The same trend continued when prokaryotes increased in size and consequently ingested their membranes along with other microorganisms resulting in the formation of markedly distinct intracellular, membrane-bound structures, marking the genesis of eukaryotic organisms. These eukaryotes then continued to digest other prokaryotes resulting in mergers between two distinct organisms with different DNA sets – animals possessing mitochondria and plants harbouring mitochondria and chloroplasts. In particular, this form of merger was 'packaged' to eventually benefit the two merging entities. The same way bacteria started taking other smaller 'packaged' ones and storing the beneficial bits, animals took 'packaged' entities, stored the beneficial ones and eliminate the harmful ones. Humans even take fortified meals, which are a form of packaging, by equipping certain foods with various micronutrients and minerals. This trip in evolutionary time captures the emergence of complexity through an increase in size by merging. Organismal Selection underscores that an increase in the scalable size is marked by an increase in unscalable complexity. Thus, throughout evolutionary history, an increase in both size and complexity establishes that an organism is still the fundamental unit of diversity.

Lovelock gives a great picture of complexity at its best and the various outcomes besides the possibilities of mergers. Take bacterial invasion, for instance. If the bacterial invasion does not interfere with the comfortable existence of the human, then the goal of the bacteria and the humans do not collide. The 'human' is, therefore, a different organism, but we cannot superficially acknowledge this at that moment in time when the merger happens. When the bacteria start increasing in size (number), there is a naturally expected increase in temperature of the human being. At this point, one is said to have a fever. The bacteria now pose a credible threat to the human. The rise in temperature is thought to be bactericidal (capable of killing bacteria), as credible a threat as any can get. The chances of escaping the threat, at times, are very few. The size of the bacteria equips it with resilience that can allow it to last long enough to spread to another host, another human being, say, when the current host sneezes or coughs. Otherwise, the bacteria die.

What of Gaia? Something similar happens; a global increase in temperature effectively called, greenhouse effect, follows an increase in 'bacteria' known as human beings. The goal of the Gaia does not match that of human beings, so raising temperatures is a means of eliminating them. We can sneeze or cough out the bacteria but the Gaia cannot, so we seek strategies of sneezing ourselves outside the Gaia, by seeking other habitable planets. Should we fail Gaia stands a chance of surviving and humans might possibly die in large numbers. It is normal for bacteria to reproduce and increase in numbers but it is not 'normal' for the human being to contain such an exponential increase in bacterial number. It is normal for humans to reproduce and increase, but it is not 'normal' for Gaia to contain such a large number. In the same way, there is little chance for bacteria to escape the human host, there is little chance for man to escape the Gaia, our host.

There are four possible outcomes: acute resolution, chronic infection, destruction of the host or, symbiosis (a lasting relationship between the host and invader). Acute resolution is likely to be the outcome as it is easy for the Gaia to eliminate human beings. Typhoons, hurricanes, glacial melting, and increasing ocean levels are few of the acute ways the Gaia can heal itself. Chronic infection implies that humanity will be present in a long time. However, given the current status of events, this is highly unlikely. A time when Gaia experienced chronic infection was at the period preceding the split of Pangaea into the Laurasia and Gondwanaland, and later into the currently existing continents. This is a form of *production* rather than *reproduction*, as I shall elaborate in the next chapter. The destruction of the host is also a possibility. Habitat destruction and invasion of species ranks high as causative factors of species destruction. Humans are doing just that, including overpopulation, overharvesting and pollution. The resilience of Gaia is dependent on the species diversity, and destroying it also destroys the diversity and hence the resilience of Gaia. It takes millions of years to establish a robust set of species. In our current trend, we might not live long to witness the rejuvenation of Gaia, after our 'acute elimination'. Symbiosis should be our goal, since it is a form of a merger, with hierarchical reciprocal returns. Ecosystems thrive on such relationships. According to OS, all organisms thrive from such relationships.

This example highlights that efficient increase in size is though hierarchies, and hierarchies evolve from the ground up. That is, hierarchies underscore the evolutionarily stable mergers. Life as we know it evolved from a single ancestor to what we presently have. Not the other way round. The prokaryote developed from another self-organizing system, a particulate system, and not the other way round. Size increase then becomes efficient by having organisms that sustain their hierarchical relationships. As previously stated, the purpose of the upper ones is to serve the purpose of the lower ones and vice versa. The purpose of the cell is to serve the purpose of the DNA. This could perhaps explain why Dawkins thinks of bodies as entities that fulfil the wishes of dictatorial genes. Hierarchies tell a different story.

The purpose of the DNA is to serve the purpose of the purine and pyrimidine rings. The purpose of those rings is to serve the purpose of those atoms that form the elements. The purpose of the atom is to serve the purpose of the proton and the electron. The purpose of the proton is to serve the purpose of the quarks. So, basically when two particles fuse, or two entities of variant organizations, there is a resultant entity with a hierarchical relationship which ensures that the involved entities postpone their respective annihilation. For them to survive they have to serve the purpose of the other. The fractal behaviour persists from the multicellular eukaryote to the particles of an atomic nucleus.

This reminds me of black holes. I consider black holes a mystery of simplistic beauty. They are so simple in their fundamental similarity with each other. The particulars known to scientists that distinguish one black hole from another is the mass, charge and momentum. With these few descriptive features, they are almost like electrons in this sense. I also consider them to be metamorphosed stars. Stars, now that we are talking about size, are large celestial bodies that would therefore have to live for long if we are to use the principle of equivalence. But since they are large, they can surpass a critical threshold where the attractive force of gravity exceeds the repulsive force generated by nuclear fusion. What follows is the star either explodes into a supernova or metamorphoses into a black hole. The black hole then variably maintains features of the organism such as its original mass while adopting new ones. For example, it has a boundary like any organism, known as the event horizon. It also has a gravitational attraction so strong, that light in it cannot escape. Light particles, therefore, form part of this organism. It allows for more mergers than fall-outs.

But even more surprising is that black holes turn out to be very small objects. It might therefore appear that they should last for a shorter time than stars because of their smaller size. However, this is not the case. Recall how fractal geometry explains how the surface area of a tennis court can fit into your chest cavity as a lung? Or how the meters long intestines fit in humans? Or how the total length of blood vessels in an adult human being is as long as a third of the distance from the earth to the moon? It all lies in the fractal geometry. Black holes display the same property to a fine degree. They are the perfect embodiment of space-filling that we know. If we are to superficially look at size, then black holes could appear small. They do so because they have highly optimised the organismal trait of space-filling, making the most of space for maximum interaction and exchange of information with particles. Since they have the masses similar to several stars, superficially, they might not be as large. Unfolded, however, they occupy a lot of space, space that stars could not do. As such, black holes are small but very, very dense. Lucky stars mutate into these heavy dense small pieces of organisms. The black hole at the centre of our galaxy is four million times the mass of the sun, but with a width of about 17 times that of the sun.[49] To my knowledge, this is the epitome of space-filling. As a result, they take billions of years for them to fizzle out, for them to die. Size, then can be appreciated in terms of space-filling as the most optimal way of increasing in size, not just by superficial assessment.

It is important to note that even though there are generalizations using the scale tool, it does not eliminate the fact that the fundamental unit of diversity is the organism. For instance, the comparison that is made using Kleiber's law is at the inter-species level and not intra-species. At the intra-species, level there is a wide latitude of differences that we attribute to organisms of that specific species. In fact, at the individual level, there are other aspects such as basal metabolic rates that are ignored, specific organs metabolism and organ system metabolism that vary, again hinting at the diversity and autonomy of organisms that gives it a dynamically stable quantity, such as the ¾ metabolic scaling that Kleiber documented. It is much like temperature regulation in homeotherms and its invariance despite the different mechanisms operating within an organism; or the fairly stable global temperature in spite of the extremes of the poles and the tropics. Introducing the ¾ power scaling was only an exercise at showing the incentive for organisms to increase in size, to reap the benefits of economies of scale.

As I conclude this chapter, I would like you to imagine a Venn diagram, comprising three circles each for Mathematics, Physics and Biology. These three circles intersect somewhere in the middle. I would like to entertain the thought that OS illustrates the possibility of these subjects intersecting in this middle region, as a consilience of sorts. How I arrived at this conclusion was by thinking about scale in terms of two great theories of physics – relativity and quantum mechanics. Relativity focuses on the larger scales while quantum mechanics focuses on the small ones. They however never quite unified like electricity and magnetism did. This has pushed some physicists into believing that one has to be correct and the other false or that they just have not found the missing piece of the puzzle.[50] I felt otherwise as my first conjecture. Why so? Each of these theories has had marvellous predictions and utility beyond physics. However, OS can offer a different perspective that physicists can consider. From the fractal behaviour of nature, the quark behaves much like you and me. The *small* quark and the *large* human being. These two entities are stable enough for us to study them. The former is efficiently studied in quantum mechanics, and the other utilizes concepts from general relativity. Only when I factored in the effect of size, which through atomic or biological symbiogenesis results in the formation of new organisms, do we get these stable systems. Stable, resilient systems stem from self-organization. Self-organization allows us to study them. The stable tripod quark system, the atom, the bacteria, the protozoa, the mammal, the fish, the bird, the insect and the fungi. These are stable systems of self-organization that resulted from size effects, particularly through mergers. At each level, we observe a new organism with emergent properties. I believe that the great divide between these two theories lies partly in self-organization. Self-organization allows for the study of the hadron, the atom, the cell, the microbe and the galaxies. Biology, mathematics and physics and are unified through this powerful concepts of size, symbiogenesis and more importantly, self-organization. Size potentially unifies the sciences!

Increasing in size, however, is not a full-proof solution. Allometric increase in the size of the human head relative to the body allows it to survive comfortably into adulthood compared to a baby born with a large head as in the case of hydrocephalus. You might have a large cranium but that might affect your balance or speed. The ability to avoid annihilation merged with the competency of annihilation speaks of the limits of size. Even if the increase in size has been tapered over generations to the one group of organisms that utilise energy efficiently, there still are constraints. We have already seen how cancer creeps in due to this need to increase in size. Similar outcomes are also seen in various medical conditions such as diabetes, obesity and other metabolic syndromes. This would then introduce us to the next concept: reproduction. In the next chapter, I hope to persuade you into understanding that reproduction could not be for furthering a gene's capability, but rather an outcome stemming as a consequence of size constraints.

CHAPTER SEVEN

From Reproduction to Production

*Cry my eyes out for days upon days, such a heavy burden placed upon me, but when you go hard your nays become yays, Yankee Stadium with Jays and Kanyes – **Fly, Nikki Minaj feat. Rihanna***

The only other provably stable relationship, other than and equal to that of nirvana (perfectly balanced cooperation), is intractable conflict. This is mathematically verifiable, as Steven Strogatz describes it in his beautiful book, *The Joy of X*.[1] It is from this literary masterpiece that I found a mathematical equivalent that supports the map we have been discussing – the organism and its universe – further highlighting the beauty of math in capturing the formal and the empirical. Empirically and thus historically, Strogatz managed to demonstrate, using diagrams, a series of conflicts that he posits, probably lead to World War I, simply because states did not want to get annihilated. Formally, by creating the mathematical framework he illustrated the sequence of these conflict-ridden relationships between countries.

Strogatz further explains that there are only two stable maps – complete cooperation, that is, a state of nirvana and its complete opposite, polarised conflict.

Analytically, OS shows why each of these relationships is a stable one – on the one hand, cooperation to form an emergent organism stable enough to avoid annihilation; on the other, polarised conflict to destroy the various mergers as in the case of bacterial diffusion to excrete waste material. What is more is that in the grand scheme the formal relationships demonstrate how stable an organism can be in a universe that is hell-bent on destroying it. This, by definition, is polarised conflict. We have mathematical reason to believe that the map is the single most stable pattern of relationship, which could explain why it is robust. While our arguments in the previous chapters might not have had solid mathematical backing, hopefully, it now does. But mathematics does much more than that. It introduces the constraints of size, which I believe is one of two reasons why reproduction happens. We can obtain this insight from Galileo.

Galileo is arguably one of the first people to capture what is now known as the square-cube law. In his book, *Dialogues Concerning Two New Sciences*, he shows that the ratio of area increase is not proportional to volume increase.[2] Take, for instance, a square of 2cm on each side, and a cube of 2cm. The area of the square is 4cm^2 while the cube is 8cm^3. If they each had sides of 3cm each, the area would be 9cm^2 and the cube would be 27cm^3. With a unit change from 2cm to 3cm, the area has increased by 5 but the volume has increased by 19. The rate of increase of the volume is much greater than that of the area. This might seem obvious but the key insight that Galileo added to the law was that the area corresponded to strength. To appreciate this, we need to take the area to volume ratio and see how it relates when scaled up:

Unit	Area (A)	Volume (V)	A:V Ratio
2	4	8	1:2
3	9	27	1:3
4	16	64	1:4
5	25	125	1:5
6	36	216	1:6

From the table, it is clear that the volume to area ratio increases when scaled up, and reduces when scaled down. If the area corresponds to the strength, then the smaller organisms, such as ants, with the smaller volume to area ratio possess much more scale-adjusted strength than an elephant. Ever since I was small, I had always wondered how kids could easily carry small children in spite of their sizes. The answer, it turned out, lay in the ratio between the surface area and their size, which gave them relatively more strength than adults. The same explanation would explain why an ant can lift up to 50 times its own weight or more but a horse cannot carry another on its back. It is not that an ant is very hard-working, but that its size grants it a lot of strength.

The square-cube law was the basis for the development of the body mass index (BMI). The BMI is a ratio that is used to classify patients or persons into different categories for easy diagnosis of obesity. The ratio, however, is flawed in the sense that it compares the body mass, a variable that is easily translated to the size and hence the volume, with the square of the height, which is the area. We have already seen how a unit change results in a disproportionate increase in volume relative to area. So, those who are tall and lean are likely to register high BMIs and the reverse for short people.[3] It does not factor in the disproportionate scalability of these two variables. A much more accurate comparison would be for the body mass and the cube of the height, which was suggested by the Swiss physician, Fritz Rohrer. What was initially known as Rohrer's Index is now known as the Corpulence Index with a much higher sensitivity than the BMI.[4] Since it factors in the very tall and the very small, it has better applicability in paediatrics and the lean individuals.

As a slight detour before we revert to its relation to reproduction, I was able to formulate an easy means of capturing the scale-adjusted weight for height. I figured that the weight of an individual should normally scale by an exponent of 2/3 the height of the individual. A person weighing 64kg as the numerator should therefore have a denominator of 16 ($64^{2/3} = 16$). As a student doctor, I tried coming up with a simple yet practical way of estimating what ought to be normal weight and scale it to a measurement that can easily be derived by the clinician. I discovered that the denominator was equal to the twice length of the middle finger in centimetres. Should one's weight be likely indicative of obesity, then the calculation from the twice the length of the middle finger should give the clinician a rough estimate of the upper-limit for normal weight of an individual. I tried it out first on my family members then on my friends and the results not only astonished me but also my friends who wondered how I was able to get accurate measurements of their middle finger. The reverse was also true, by getting the length of their middle finger, I could estimate their weight. I, however, do not have enough friends in other countries to test this hypothesis. Seeing that I was bound to face a lot of rejection for my idea in various journals as is typical in the research world, I figured why not do it here where I have no such restrictions. Thus, the rule is: as far as obesity is concerned, for one to be classified as a normal weighted individual (not obese), the weight in kilograms should scale to an upper limit of twice the length of their middle finger in centimetres. I call it the 'middle-finger' rule. Now, back to size constraints.

Scale explains why the elephant and the mouse are different. The mouse would lose much more heat than the elephant due to much higher surface area to volume ratio. The elephant would then utilise the space-filling property to increase its surface area (large ears, large, padded feet) and still reap the benefits of the increase in size. How, then is reproduction related to the strains of size? Let me answer this right here and seek to explore why the answer is valid throughout this chapter. The strains of an organism due to its size expose it to the option of reproduction, which if we recall, is a secondary means of increasing in size in biological organisms. What this means is that once an organism cannot increase in size, it seeks other means for which it can increase in size all the more and evolution has taught us that this is largely through reproduction.

It differs with the idea of Natural Selection which states that a 'living' organism will have reproductive capability but in retrospect, I find that it might not adequately explain *why* it should reproduce. Natural Selection asserts that the organism or the gene that reproduces fast with the highest fidelity should have the greatest chance at being the most fitting in many future unforeseen conditions. It largely explains *how* this works but it does not adequately explain *why*. Borrowing from the inherent incentive that organisms have to increase in size, OS reiterates that first, organisms will consider increasing in size to delay annihilation and enhance their efficiency. This happens through the natural process of physiological growth – the primary process of increasing in size. To my knowledge, the only way that this can happen is through efficient mergers. Secondly, it indirectly states that constraints of increasing in size are enough to push an organism to consider reproduction as the only option for the emergence of an even larger organism. This insight dawned on me following the square-cube law when I tried relating it to the sandpile model.

Surface area translates to two things. First, the strength of the object (organism) bearing the weight. Second, it sets the limit for extent to which an object can increase in size. This principle, among others such as geometry, guides engineers and architects when building support structures such as beams and designing buildings.[5] I argue that the same principle applies to the sandpile model. In a sandpile, a grain of sand is dropped from a single immovable drop-point onto a flat immovable surface, which ends up forming a sand hillock that over time reaches a critical point where it can no longer go any higher. An avalanche ensues. Other smaller hillocks start forming around it, but even then, these are also constrained by their square grid areas.[6] Over time, small hillocks merge and in the process, generate other smaller hillocks. This model is simple and replicable enough to show just how the state of a sandpile can result in the creation of an intricate ordered pattern of big and small albeit similar sandpiles. It was then that I posited that reproduction could just be a by-product of mathematical and physical laws. While Natural Selection states that organisms reproduce so that the genes can win the evolutionary lottery, do they know about this lottery when they have an entire universe at its neck? Maybe. Organismal Selection asserts that it is size limits set by physical constraints which are further predicted by mathematical laws result in reproduction. Absurd? Maybe. Then again, it might not. Here's how we can do away with the supposed absurdity.

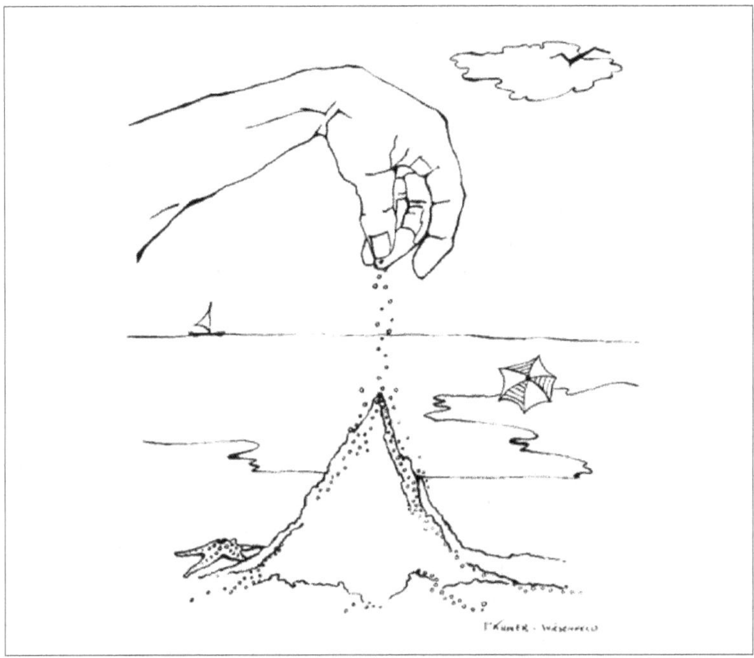

An illustration of a sandpile increasing in size resulting in the process of production. Several avalanches result in the formation of small hillocks on the side.

A keen look at the small hillocks that form on the side of the large sandpile display a morphological similarity to the original sandpile. It is as if they are the offspring of the original sandpile. To me, this rang home as reproduction. The mother sandpile was able to *produce* another version of itself, a much smaller version of itself at that. The smaller version would also grow like the parent sandpile, through mergers with other grains of sand. It was *reproduction*. Even more distinct is that the sandpile model displays features of self-organization. Firstly, there appears to be spontaneous emergence of the sandpile and an unpredictable avalanche of smaller sandpiles. Second, there is the feature of power laws that emerge from measuring the sizes of the hillocks. Each also is resistant to perturbations produced by the falling grains in relation to the hillock size. A single grain would align with the physical laws such as the law of gravity. This is an evident merger among the particles, even though biology or chemistry does not prefer us to look at it that way. Self-organization would nevertheless, end up creating a mother sandpile and baby sandpile.

The only person, to my knowledge, who considered a sandpile to be an organism is Per Bak. Together with Chao Tang and Kurt Wiesenfeld, Bak developed the idea of self-organized criticality. I am, however, not sure whether his colleagues also equated a sandpile to an organism. Nevertheless, self-organized criticality shows how systems would organize to critical states that would then result in avalanches, what OS would regard as production. The sandpile would continue to increase up to a particular critical state, after which an avalanche would be triggered. The wide applicability of self-organized systems to the point of criticality then led me to consider the vast similarity between organisms as described in Biology and as OS describes them. It is the same principle of self-organized criticality that persuaded me to strongly endorse the punctuated equilibrium theory of evolution. It would also describe the stock market, the traffic jams in cities, the evolutionary catastrophes, solar flares and even pulsars in near galaxies. All this is beautifully captured in Per Bak's book, *How Nature Works*.

Before criticality is achieved, firstly, self-organization. Self-organization manifests itself through stability and hence with the capability of generating system states far from equilibrium. Once a mother sandpile cannot contain more particles, it self-maintains. It can even remove some particles to accommodate new ones. It remains organizationally closed but thermodynamically open. Initially, the mother sandpile would grow in size through particulate symbiogenesis, that is, through physical laws. Eventually, it would end up increasing in size and create a new pattern every time of self-similarity by producing and reproducing smaller versions of itself, each similar albeit one more unique than the other. If these sand piles can be considered organisms, each is unique and diverse. Each occupies its own spatio-temporal niche. Each has its unique mass. For any new particle to disrupt a sand pile, it is only possible through the kinetic energy that the incoming particle wields, enough to dislodge the gravitational attraction and friction among the particles in the sand pile. In short, the sandpile attempts to avoid annihilation. Each sandpile *produces*, much like the bacteria or yeast, through the constraints set by physical laws or past the point of self-organized criticality. Reproduction, then, might have been an evolved strategy that organisms exploited due to the constraints of increasing size.What of the biological creatures? West showed in his theory that for a young, growing organism, there is much more energy available for growth than maintenance. The trade-off results in increasing growth. It plateaus at the juncture where the energy needed for maintenance tips that needed for growth. Physiological growth then ceases. This is the primary means by which organisms increase in size. The organism then works to maintain itself, an exercise that it does only imperfectly. Over time, senescence kicks in. Firstly, the incentive for increasing in size is evident from the very instant the organism's existence is registered. There is a natural tendency towards growth that is hardly ever stunted. The baby will grow into an adult and the egg will hatch and grow into adult versions of an insect, snake or bird. Even the baby kangaroo will continue to grow in the parent's pouch. This; however, happens until the limit has been hit, and the organism then focuses on maintaining this size. In this sense, it tries to avoid the annihilation that comes with a constant effort at maintenance. At this point, the organism acquires *similar* traits as those of any particle that simply tries to exist, whether living or non-living. It does so by seeking alternative ways of efficiently doing so, which has, for as long as it has been living, been through increasing in size. It can no longer do this through the merges it made with other particles by ingestion, the principle process through which it increased in size physiologically. It has to seek another way. This, I wager, is how reproduction began. Additionally, the concerted efforts, all in ultimate futility, introduce the other facet about reproduction – satisficing.

Satisficing is a term that was introduced by the polymath, Herbert Simon when he was studying the kind of economic choices that people made.[7] He noticed that people often settled for the most convenient rather than the most optimal option. This was often due to the insufficient information or the tendency to minimise energy, which in this case is the energy used in thinking. For an organism that is learning how to adapt in its universe, the information can never be perfect even if it has the most ideal way of thinking. Using the Popperian schema and the robust map, we can be certain that organisms are always satisficing. They will not have complete information neither will they want to waste all their energy going through all the options. They would rather and more often take the gamble and live (and/or die) with the repercussions. With this in mind, we can go back to the very first ancestor, the famously cited organism that started what has grown into a very rich field of evolution.

The common ancestor, the very first one, who by current standards we have to call 'living', was bound to reach a limit, where its maintenance could not keep up with its growth. It was bound to reach its size constraint. I like to think that once this point was reached, the organism contemplated that in the short time it is still 'alive', whatever that means, it could still salvage itself. If we yield to what biology has taught us, the ancestor would have considered producing another version of itself. It would do so through the process of reproduction, the asexual type. The untoward outcome with this situation, much in line with the Popperian schema, is that our very first ancestor thought that it was reproducing itself. The sad reality was that the map was and still is invariant – there will always be an organism and its universe. Thus, that which it 'reproduced' was not itself. Rather, it *produced* another organism. Unbeknownst to it, through satisficing, its efforts were an exercise in *producing* another organism, not reproduction.

The term *reproduction* is a misnomer but biologically it means bringing forth another life-form. After *production* there will be two organisms, each unique and with their own universe. Even when we yield to the theories of evolution that insist on *reproduction*, the organisms produced are unique and diverse just by mere existence. Why? Because organisms are satisficers. But if they are satisficers, does it not mean that the strategies will eventually change over time? Why is reproduction still present after billions of years?

Remember hierarchies start from the ground up. It will start with the first organism who will *produce* another, in some bold albeit imperfect feat at *reproduction*. It will then form another. If these organisms can communicate, then we have another organism formed since there is a merger (communication or understanding) between them. This emergent organism is much bigger than the two original ancestors. The incentive for size is evident. The first organism will serve the interest of the second by even manufacturing repeats of its DNA and try sharing it with its other 'clone'. Yet a clone is only a clone at best, genetically but a totally different organism. The second one is serving the first by being around since it is getting uncalled for support from this giant organism, its parent. On the other hand, the first could be thinking that it is serving itself, its 'reproduced' version. It could also be that the first organism is aware that it might be different from its produced form, but it tries to foster a bond with the second one. This relationship continues to serve the emergent organism that stemmed from the merger between these two ancestor organisms. Herbert Simon would insist that the two organisms are operating from points of *bounded rationality*. Each does not know much about the other and their efforts are sub-optimal. The strategies satisfy and are deemed efficient by any of the two organisms; they suffice for the moment. Thus, each of these organisms satisfices. It leads to more, 'reproduction' as each reaches their size limit or approaches their size constraints.

Reproduction then continues as the most evolutionarily stable strategy – it allows for an exponential increase in size for organisms. After productions, forged mergers create even bigger organisms. Organismal Selection asserts that this strategy of 'reproduction' is merely a step in bypassing the individual constraints of increasing in size. Exponential growth is a much faster and stable way of increasing in size besides symbiogenesis or the physiological process of growth. In short, the secondary means of increasing in size is more stable than the primary means, it's an evolutionarily stable strategy. If you are not aware of this by now, Organismal Selection has explained, albeit differently, a crucial facet of Natural Selection.The rise of 'reproduction' marked the beginning of new organisms that could increase in size exponentially and they have never looked back since then. It was much better and superior to just growth from childhood to adulthood. There was no cap between growth and maintenance as seen in the physiological processes of growth. There was no senescence. 'Reproduction' turned the plateau phase that marked the balancing phase between growth and maintenance into an exponential increase. You simply have to marvel at the organism's obsession with size. Just look at sea creatures, the largest 'living' creatures known to man. They have the advantage of buoyancy offered by the large water bodies. These creatures increased in size past the strain experienced by terrestrial animals that lack this buoyant support. However, even then, size still has a limit as maintenance gradually becomes the primary focus rather than physiological growth when these creatures reach adulthood. This size constraint would then result in *production* rather than *reproduction*. The offspring only appears to look like the parent, hence the idea behind 'reproduction', but size clarifies some aspects about this process.

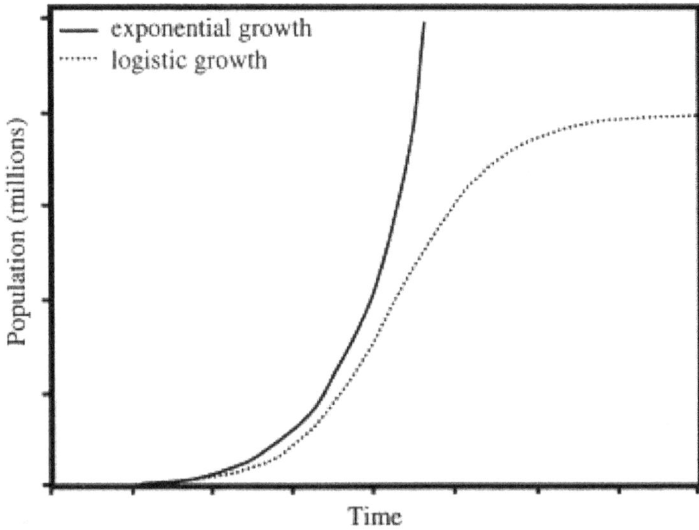

Graph showing the similarities and differences between an exponential and logistical growth. Exponential growth is incompatible with the fitness of resources. Once the carrying capacity is exceeded, what is observed is a logistical regression or logistical 'growth'. The mid-range of the curves is the range where organisms and populations increase in size through mergers. The point where the curve plateaus is the point where maintenance happens, when size reaches its structural limit. It is the adult version of an individual organism such as the adult housefly of frog.

What this would mean is that size dictates to a great extent the production of new organisms. We have already seen this in the example of bacteria and waste products. The initial amount of waste is tolerable but as the waste increases in size, it poses a threat to the bacteria, and it is released, and even then, not completely. The much that cannot be tolerated is released and a tolerable fraction remains. The released waste is a process in the production of a new organism(s) even though with reference to the bacteria it is considered waste material. My argument is that this process also applies to what we have always known as reproduction. Size constraints lead to the production of a similar organism, an offspring. It would mean that the mother and the unborn child are a single entity before the physical separation. Parturition later happens due to the constraints realised by the parent due to the increasing size of the unborn baby. It is different from the production process of releasing waste products since the born baby fosters a bond between it and the parent. Waste material does not. While both processes are stimulated by size constraints, one loses meaningful connection with the parent (according to OS) while the other fosters it. We can trace how this happens.

Using the example of human beings, which is my specialty, the formation of a new organism (zygote) happens after fertilization, what I like to call another form of symbiogenesis after the ovum and the sperm genomes merge.[8] What happens then is the movement of the zygote to the uterus where it implants. The first indication that the mother and the developing embryo are a single merged entity can be acknowledged through the process of implantation. At this stage, there is direct communication and feeding from the parent to the zygote. As small as it is, it can consume from the readily available nutrients at the epithelial lining of the uterus (endometrium).[9] As it becomes bigger it gradually asserts its autonomy. The mother also asserts its autonomy by the facilitation of a process known as decidualization. While decidualization endows the growing foetus with broken-down material for nutrition, its primary goal is to form a barrier that prevents attacks from the immune system of the parent.[10, 11] Decidualization forges a merger that allows for both to co-exist. In the process, the mother produces a lot of blood and manufactures a lot of food for the developing foetus to the point that the mother can even acquire diabetes. This is the other evidence that the two are a single entity.

This quiescent stage that extends from fertilisation to parturition is known as the gestation period. The developing foetus, therefore, continues to increase in size up to that point where it poses a risk to the mother. At times, the threat can still be present even before the birth of the baby. However, in an otherwise normal pregnancy, size leads to the stretching of uterine muscle and stimulates the birth process, through powerful contractions.[12, 13, 14] Size further contributes in part to the 'breaking of water'. It could even explain why labour is as intense as it is, where the parent's existence is at risk. The mother has to deliver fast or she risks losing herself and a component of itself. A progressive increase in size approaches constraints that eventually lead to production (what biologists calls reproduction).

The birth process does not happen after the fusion, that is, fertilization, nor even after the cell has divided several times. It happens when the foetus is *large enough* to pose a threat to the parent; when the mother approaches her size constraint. It is similar to the sandpile model, where the mother sandpile increases to the point where it cannot contain more, resulting in the production of smaller but similar sandpiles next to it. The baby sandpile similarity in appearance can give the impression that it is the offspring, as if it has inherited some features of the parent. Here we have what biologists can call an inanimate aggregation of particles but it, nevertheless, results in an 'offspring' sandpile similar to the 'mother' sandpile, and one that grows just like the mother. There is no inheritance of genes in the Mendelian sense but there is self-organization. Self-organization allows the parent organism to act in good time to prevent itself from annihilation having increased in size thus far. For the infertile organisms that cannot self-organize in this sense, that is, increase in size through 'reproduction', there are alternative means such as creating mergers through language, for instance. For inanimate particles it is a means that shows how self-organization is present even in the 'inanimate' systems and that they bear similarities to those of genetic inheritance in 'living' systems. I would rather call all these processes production that follows from the constraints of size, rather than reproduction, which strictly speaking, can never happen.

Once this idea is set it would explain numerous processes observable in nature. The dispersal of seeds from overhanging trees, branches and flowers, falling of leaves in different seasons or drought, transpiration, respiration, excretion and 'reproduction'. These become easily appreciated as production processes that stem from constraints of size. Even the reproduction process distinct in yeast, budding, can easily be described by the approach of an organism to its size constraint.[15] It is very similar to a cloud, which increases in size but after reaching criticality, produces rain. It takes us back to what we said in the previous chapter about size – it gives an organism just enough time to 'reproduce'. What we can now say is that size exposes the limits of an organism, which lets it release some of its contents to avoid annihilation. The bundle of contents released could thereafter bear little use to the producer (as in the case of excretory waste) or meaningful use (as in the case of offspring).

I would like to think that if reproduction, as purported by the conventional evolutionary theory of Natural Selection, were to ensure genes continue existing in the gene pool, then evolution ought to have found a way of allowing shorter and shorter gestation periods among different species. A shorter gestation, before the child became big enough to pose a threat to the parent, would have secured many parents from death. The parent could then have found ways of caring for it in case such a merger comes in handy in the future. Secondly, over time, organisms would have earlier and earlier reproduction periods. What I mean by this is that if an adult human being achieved mature reproductive capability at 24 years, then evolution would have worked such that after subsequent generations this reproductive capability would have come much earlier than 24. This slight change would have given a group of organisms an edge over the others even before becoming 'adults'. However, it turns out that species have to become adults first before they 'reproduce'. Couple the two and you find that an organism that would evolve to have shorter gestations and consequently smaller babies, and attain reproductive maturity earlier, would have a greater chance of success than those on the other extreme end. However, biological organisms release offspring when they are significantly large, large enough to pose threats to the parents during this reproductive stage, especially among vertebrates.

The most fitting species are indeed the ones that will survive. What we add to this description is that the living organism that is *most sensitive to these size effects and acts fast* will survive. The hypothesis that explains these phenomena has to discuss the effects of size. From OS, we hazard a guess, one that extends into particulate organisms. In short, the organism is ever driven to increase in size, but size exposes limits. The sandpile will produce smaller piles and the mother will produce small offspring due to size constraints. It was Donella Meadows, in her book, who explained to me how growth exposes limits, even when it is the one thing which is always emphasized the world over. Under the pressure of these constraints, if the organism is efficient enough to produce, the result is an offspring. If it cannot cope, it results in the collapse of the entire system. The stage is thus set for revisiting the idea we had about cancers.

Cancer, as we have already seen, can be viewed as a natural consequence stemming from an organism's obsession with size. We also noted that due to this obsession the mouse had more *diagnosed* tumours than the elephant or even the blue whale. Different reasons have been churned in the field of oncology for this strange phenomenon. The first is gene-centred, in which the larger mammals have been reported to have more regulatory genes such as the famous p-53 and Rb genes.[16] Caulin and Maley have also synthesised other reasons such as enhanced response to cell-death, different genetic lineages and cellular architecture. However, what I like to think holds more ground with our theory is the hypercancer hypothesis. A hypercancer is a cancer of a cancer.[17] It forms when the cancer has become so big that it creates another mutation that forms another genetically distinct cancer. Why? The replication of the genome is never perfect. Therefore, after a series of continuous replications, the original cancer cell is bound to result in another cancerous cell with a different genetic picture. The larger the animal, the more room for replication and, hence, an increased chance of forming of brother and sister cancers. In technical jargon, from one clonal expansion we arrive at more divergence and production of other clonal species. Like the tree of life, we have a vertical evolution and speciation.

Cancer, therefore, is an excellent candidate in explaining evolution. As we have reiterated, the tendency to increase in size exposes the limits of the organism. Even in cancer, these large mammals expose the cancer to its limits, which results in the *production* of more cancer. One clonal expansion, given enough space and time, produces another. Since each is an autonomous entity, gaining resources from a common source, they will balance each other and appear as if there are no debilitating effects of the organisms (disease-causing cancer) to the host (the larger organism bearing the primary and sibling cancers). In a concept in systems known as balancing feedback, the original cancer will limit the resource for the new cancer, and the new cancer will do the same for the original cancer. They create a balance that ends up not hurting the blue whale or the elephant.

It is my belief that every living organism contains cancerous cells, or, if not, cells that display cancer-like behaviour. However, each cancerous organism exposes itself to its limits, the available yet finite resources, creating a competition with other cancers that yielded a balance. The autonomy of each cancerous organism creates a dynamic stability. For the mice, they do not have enough space for cancers to reach such limits as to produce sister and brother cancers. Even more, since they are small, they do not have enough time to develop hypercancers. Small animals do not have enough *space* nor *time* to produce hypercancers. Nevertheless, the incentive to increasing in size leads these small organisms to continuously incline in this direction, resulting in a greater incidence of tumours in smaller mammals than bigger ones. I believe this solves the Peto's paradox. Taking the idea of clonal expansion as the emergence of speciation – or organism, as OS puts it – the same phenomenon should explain the widely documented means of speciation (such as allopatric and sympatric) and even adaptive radiation.

Understanding how each organism can be detected can then be used in the control of diseases from a systems perspective. The strong test of organism presence is subjecting it to a credible form of threat. Subjecting the cancerous organism to chemotherapy will, over time, lead to resistance. Yet, cancer displays features of evolution just like any other recorded species. While therapy targets completely eliminating *diagnosed* cancers, there are smaller *undiagnosed* ones that will occupy this ecological niche created in a feat of adaptive radiation (when species spreads to occupy a wide variety of ecological niche). Each of these organisms yields to the incentive of increasing in size, and are balanced out by other equally competing cancerous organisms. Learning to control rather than cure cancer can be a better means of culturing a symbiotic relationship between cancerous and non-cancerous organisms. Detecting the highly resistant from the benign cancers would then guide whether we should 'cure' the cancer, even though there may be no such thing, or controlling it.

Living within Limits

Size explains production, excretion, shedding, pollination and now cancer. Furthermore, through cancer, it also describes speciation. It also introduces a rather peculiar concept, that is, invariance. The beauty of size, if I might not have explicitly said it, is how it introduces the ability to scale up or scale down, with certain aspects remaining invariant. We can list them up. Firstly, I have shown how the map remains invariant, regardless of the size of the organism. Secondly, the tendency to avoid annihilation, is invariant, regardless of the size. Thirdly, the definition of an organism remains the same. Fourthly, the ability to detect the presence of an organism, by introducing a form of credible threat, is also invariable. There is another fifth one, which is, tendency towards size exposes an organism to its limits. This is an inevitable consequence in all systems, biological or otherwise.

Donella Meadows once stressed that growth exposes the layers upon layers of limits in systems. Per Bak specifies that systems self-organize to criticality. In OS, we simply play with the words to say that size introduces constraints. From these limits or constraints an organism can strive to maintain its organizational pattern by *producing* rather than *reproducing*. Since scale allows for invariance, we are also able to maintain these five invariant properties when we shift to societies. First stop, Robin Dunbar.

Robin Dunbar is famous for his Dunbar number hypothesis by correlating the primate brain size with the maximum number of stable relationships that can be developed in a society. The number dabbles at approximately 150 members.[18] Once the numbers begin to swell past that, it becomes difficult to maintain stable relationships. Malcolm Gladwell explained how these numbers have been maintained in stable community systems from early human populations to certain branches of various companies.[19]

Using the famous game of Prisoner's dilemma one can deduce that these stable relationships are built from a state of continuous interaction. If I know I will see my neighbour tomorrow there is little to gain from defaulting from some expected duty. If I shirk my duty, my reputation in that community is destroyed. I can even get annihilated when they opt to kill me. Thus, the members of a community can foster stable hierarchical relationships through repetition of duties, continuous, regular interaction and the need for maintenance of reputation.

Since there is coordination of behaviour, where my actions can be reciprocated and avoidance of tarnishing one's reputation is evident, there is communication between the members. This constitutes a merger. Therefore, in societies, there is an organism present that is made from the various combinations of mergers between and among the members. To test if the organism is present a form of credible threat such as a raid, war or even famine forces the community to seek ways of enduring through nature's painful and often unexpected jabs. The strategies they resort to determine if it survives or not. Enter the role of the elderly in societies that prove pivotal at such crucial times. What's more is that during times of bountiful harvests and peace, the community increases in size. However, the one invariant principle is that size grows to expose its constraints. Bountiful harvests, then, are not as glorious as they appear. They foretell the future instability of the community, reiterating the role of the universe in wielding ingenious strategies of annihilation.

As the number approaches 150, according to Dunbar, the stable relationships approach criticality. Past the number or more accurately, the critical point, one can risk shirking duties and their reputation can still be intact. The society's limit results in *production*. It starts producing defectors rather than co-operators. Defectors are a new breed of organisms, just like hypercancers. Taking cooperation and defection as traits, the clonal expansion of co-operators could not replicate the trait anymore. It then worked at maintaining itself after attaining its limit. A new clonal expansion of defectors starts to form. They begin seeing benefits from defecting than from cooperating. Defection then becomes appealing, earning them greater returns than cooperation. It is a new organism because having realised the benefits of defection, cooperation is a threat. Cooperation denies defectors favourable returns or relatively large gains. Selfishness steadily grows from the initially stable conditions of cooperation.

This simple illustration does more than indicate the emergence of new organisms through defection. It also sheds light on the initial ancestors. Had the first living organism 'reproduced', it should have coordinated its actions to suit what it thought was a production of itself – a 'reproduction' of itself. The second organism had to coordinate its behaviour for it to understand the first organism. As for membrane-bound vesicles, which likely preceded prokaryotes, coordination of action allowed selective inflow of material, setting the stage for the evolution of complexity. If you will, the selective permeability, likely unknown to them (or us), by definition is indicative of communication and hence a form of merger. In the case of prokaryotes, the second organism had to coordinate its behaviour by taking in pieces of DNA that the first one had produced and 'shared' by releasing it into its immediate space. For there to be such coordinated behaviour, constituting another level of communication between the two, it presents evidence that there exists another merger. Since coordinated behaviour is the hallmark of communication, the original ancestors could have started formulating seeds of communication by this means. They would produce copies of a DNA and throw it at the other 'original copy'. The recipient had to coordinate itself to accept the piece thrown at it. This continuous behaviour constituted a merger. Hence the original ancestor expedited not only the emergence of the third organism but also the development of communication. However, recall that coordination involves movement. This would happen the moment the first two particles, inanimate by current standards, began to move in phase and hence coordinate. This could actually be the initial stages of communication.

Such mergers demonstrate much more than what a gene-centred view or even Natural Selection made us believe. That is, there actually was a speeded-up process from one organism, the original ancestor, to a third organism, after the merger with the second one. What this further means is that cooperation speeds up the emergence of defection. An alternative way of looking at it was that cooperation through size increase speeded up speciation. The originals are bound to increase in number, just like any society, but faster than expected because of the mergers generated. Consider a pairwise link, a merger as simple as that between two people. The number of links can then be generated using the formula:

$$\frac{\text{(Number of particles)} \times \text{(Number of particles -1)}}{2}$$

If there are two particles, the pairwise link is $(2\times1)/2 = 1$. If there are three particles, $(3\times2)/2 = 3$ and if four $(4\times3)/2=6$. If they are five, $(5\times4)/2=10$. The number of pairwise links increase much more than the particles. This is similar to the idea that Kauffman had about particles with general catalytic activity resulting in more catalysis than complex molecules. Taking the initial stable cooperation that marks increasing organismal size, the mergers demonstrate how fast organisms are generated and how quickly the size constraint is reached. Past a certain threshold, defectors are produced or, if you will, new species get formed. According to OS, new organisms get formed at this point. This is the general process an organism undergoes – it increases in size, approaches its limit, and then produces another in order to maintain its existing condition, which is 'live within its limits'. In the words of Per Bak, the system self-organizes to the point of criticality.

Before we further explore this idea of size constraints, let me take you back to the inverse law of large numbers. From the thought experiment of balls with the ability to merge confined inside a room, we concluded that with each merger there is the increase in the diversity potential in the room. The balls represented the organisms and the room represented the shared idea of a universe. We then used this to explain how the Cambrian explosion could have happened over 540 million years ago. Having reached this far in the book, we can fine-tune our initial explanation bearing in mind our argument about catalysis. For molecules to continue existing as they do, they have the baseline catalytic property. Catalysis is also a form of a merger, as we have previously discussed. Consider that **A** and **B** can spontaneously combine to form **AB** in the morning and **BA** in the evening. Now, assume that **AB** catalyses the reaction of **BA** and vice versa. This is an example of an autocatalytic set. Such a set has the potential of stimulating each other's reaction to the point that it would not need to just happen either in the morning or in the evening. The autocatalytic set can evolve to become unperturbed by daily variations. As mergers increase exponentially relative to the organisms, what you observe is an explosion of possibilities. Besides the increase in diversity potential, you literally experience an increase in diversity. Mergers could explain the Cambrian explosion following the idea of criticality. That is, mergers increase much more than organisms, which increases organisms all the more in a critical cycle. That being said, we can get back to the size constraints and how they have been surmounted by a certain group of organisms.

Eusociality

There are special groups that have found an ingenious way of eliminating these size constraints. As a result, they can continue to increase in size with, to our knowledge, much more unbounded size constraints. This efficiency is seen in eusocial organisms. The prefix *eu* is Greek for 'good' or 'true', so eusocial means the good or truthful social interactions between organisms. They represent the highest known level of organization of sociality – *eusociality*. Termites, ants, bees and even naked mole rats, are considered eusocial and, hence, they have occasionally been regarded as superorganisms.[20]

Eusocial groups have only one person from the entire colony, the queen, who is chosen to 'reproduce'. The primary role of the queen is to give birth and feed. The rest work, raise the young and protect the colony and the queen. We can make a different interpretation of this. Eusocial colonies have an incentive to increase in size. Rather than increase by letting *everyone* 'reproduce', the size can be increased by letting a *single* person 'reproduce'. This strategy has its benefits. For instance, there is an increase in size with very few reports of death through the process of 'reproduction', which is seen in very many social species, humans included.[21] I also consider these eusocial groups to be organisms that have found ways of bypassing the constraints of increasing size by eliminating the multiple possible defections. Multiple defections arise whenever more than one person 'reproduces', just like cancers and hypercancers. However, having a single person that does the 'reproduction' links everyone together, with limited temptation for defection. Those that defect would have a difficult time increasing in size since they are variably sterile. The rewards for cooperating and unlike other organisms increasing in size, it lowers the temptation for defecting. The colony then behaves like the *eu-organism*, increasing in size but with minimal defection. As an example, due to this effective means of increasing in size, the ant eu-organism can live for up to 15 years! Size does really introduce efficiency and delays annihilation. The *eu-organism* is one of the most efficient organisms recorded in evolutionary records, and OS shows why. The prevalent hypothesis suggested by J.B.S Haldane and formalised by W.D. Halmilton, has proven insufficient in explaining the existence of eu-organisms. Other leading researchers have thus offered an alternative explanation.[22] With the benefits of size, we tend to explain much more. For instance, the males in a bee community tend to display features of an organism just as much as the females. The males are organizationally closed, but open to the idea of increasing in size through the queen. On the other hand, the females are organizationally closed by their various features, but open to the idea of increasing in size through the same queen. The role of the queen therefore serves the female and the male bees. The roles of the female and the male bees then serves the queen. These groups end up forming superorganisms in this measure.

What OS explains is not only why organisms increase in size but why they evolve towards eusociality. Eusociality has been noted to evolve independently at least 17 times in evolutionary record. This is only among the species that have been studied. In terms of biomass, the ants account for more than 50% of all the known insect species and are more than four times greater than all non-human vertebrates combined. Eusocial species, what we have called *eu-organisms* really are dominant. In addition, we can test if indeed they are organisms. Posing a credible threat, especially threatening the queen, is enough to have the entire colony attack. I also have to insist that the queens do not reproduce – they *produce*. With every production there is a unique organism joining the colony. The uniqueness is relevant in ensuring the resilience of the *eu-organism* as has been detected multiple findings such as the stabilising of warmth of insect nests among honeybees and in enhancing disease resistance in some ant species.

Size Effects

The universal obsession with size is but a short step from finding how organisms are extremely sensitive to size effects. Here is an example of what I mean by size effects. Temperature is a size effect that stems from several small particles moving about in a confined space. You cannot get the temperature of a single particle; they have to be several and moving in a confined space to know the temperature of *that* space. The human body is sensitive to the size effects of particles registered as temperature. The core of the body can start shivering during cold temperatures and the skin sweat intensively during hot temperatures. Even organ systems are sensitive to size effects. The circulatory system is extremely sensitive to pressure changes; pressure is a size effect. When the blood pressure gets high certain balancing effects are triggered. The heart, for instance, starts producing a regulatory peptide to increase the rate of urination until the pressure gets low, then it stops producing it.[23] Nerve cells too respond to the size effects. Once the neurotransmitters accumulate past the necessary critical point they fire. It is famously known as the all-or-none principle of nerve conduction.[24] Memory, intelligence or even understanding for a complex organism can, therefore, not be directly atomised to a solitary particle. It has to be a size effect that leads to excitation or inhibition of nerve action. When the sensitivity to size effects diminishes, diseases crop in. Particles accumulate in cells resulting in cellular death and end-organ-damage; circulatory system is pushed to the brink of instability where it cannot execute corrective compensatory actions; nerves start misfiring or do not fire altogether.[25, 26] Sensitivity to size effects is crucial to the survival of organisms and in particular, living organisms.

I can extend the same sensitivity to the process of birth. For there to be an acute sensitivity to size the parent and the child have to be identified as a single entity. I have already explained how this happens from implantation through to the process of birth. Resorting to a little bit of detail might clarify this some more. A good place to start us off is the circulatory system.

It takes a while for the blood vessels to form during the process of implantation. However, once the blood vessels have already formed, the transmission of nutrients follows a unique path in the developing foetus. It is so unique that we give it a different name – foetal circulation.[27] Since primary school, I have always known that with various exceptions, oxygen and nutrients move from the arteries to capillaries and through to cells while carbon dioxide and waste products produced by the same cells are released through the same capillaries and transported in veins to complete the circulation. Remember that nutrients, oxygen, carbon dioxide and wastes are effectively released through size effects. This, however, is not the gist of this particular argument. It is just a reminder to show you how the organ systems are effectively functioning through sensitivity to size effects.

The gist of our argument is that the foetus is able to coordinate its circulation to match that of the mother. Nutrients will move from the arterial system of the mother, through a capillary bed found in the placenta then into a venous system that distributes oxygen and nutrients to the foetus. The lungs of the foetus, at this point, do not serve the same purpose as those of the mother. The lungs are yet to take in and release air, unlike those of the mother. There appears to be a coordination in the maternal and foetal physiology. Coordination of behaviour translates to a single system comprising the mother and the developing foetus. In fact, the placenta is even formed by cells from both the developing child and the mother.[28]

The mother would then continue to feed this growing bit of itself, until she approaches her size constraint. She has to be sensitive to the changing size of the developing child – to the size effect – for it to survive. It then works out an elaborate and efficient process that breaks the sac holding the amniotic fluid that houses the baby. This would be the first sign that in the next couple of hours it might just be the baby's day out. Labour soon begins. Since the urge to release this growing entity poses a threat to her, it is as efficient as it is painful. What happens next is a beautiful display of self-organization.

First, the child takes its first breath. In this way, the infant bypasses the limits exposed by growing inside the mother. The baby can then continue to increase in size, unencumbered by the small space of the uterus. Thereafter, the two entities work out a coordinated relationship, so much so that physical separation through the cutting of the umbilical cord does not separate them. The mother reorganizes its system to synthesize a product it initially could hardly do – it produces milk.[29] The child has to behave in a way that coordinates the actions done by the mother, by all the means it wields. Placed on the belly of the mother it is able to notice the smell produced by the mother and arch as if intending to move towards the breast. Before this, the child was temporarily separated from the mother when the umbilical cord was cut. It, therefore, had to self-organize and reformulate the merger from a physical one to a behavioural one. Lastly, the infant seals all the shunts it had when it depended on the mother's circulation. In the process, its circulation changes from the foetal circulation to the normal human circulation. The concerted efforts of the child and the mother, then seek to develop a new merger, fostered and nurtured through lactation.

While the mother is lactating, there is the production of a hormone that facilitates two roles. First is the extraction of the placenta, which again, due to size effects of increased circulation to the placenta, poses a threat to the mother.[30] The other role of this hormone is to give the mother feelings of joy at the new behavioural merger it has formed.[31] This hormone is oxytocin. Throughout the existence of the embryo through to the live infant, there was a physical manifestation of a merger (placenta and umbilical cord formation), a behavioural merger (suck reflex and lactation tendencies) and one among the very many molecular signs (such as oxytocin production) in this system, highlighting that they are not just two entities, but a single one. They are a single organism. Their coordinated actions signalled communication. When the umbilical cord was cut it was the ready threat that was resisted by forming and strengthening another bond, the success of which is pleasurable sensation produced by oxytocin. The same hormone is also partly responsible for aggression following attempts of causing harm to the baby.[31] Like the covalent bond that needs to be broken between two carbon particles, energy is also needed to break the mother from the child.

Let us then revise some of the features of self-organization that are evident through this process. Firstly, self-organization is a manifestation of order. Spontaneous order. There is the almost fluid conversion from the different cell stages after fertilization, to the placenta that has cells from both organisms which eventually forms different types of circulation. There is even the spontaneous conversion from foetal circulation to normal circulation after birth. Secondly, there is coherence. This much is seen by the means with which the foetus and born baby coordinates its actions with that of the mother to form the various forms of mergers. The other is power laws. Perturbations result in resistance. Even cutting the cord does not break the bond between the two. This is spontaneous order from the molecular level to the individual level. It then espouses the role of hierarchies. The role of the mother serves the child, and the role of the child, in having the potential for greater physiological growth than the mother, serves the mother. This is a stable hierarchical relationship. However, the placenta does not. After birth, the placenta continues shunting blood from the mother but has no other place to distribute it – the baby has already been born. Therefore, the role of the placenta does not serve the mother, while the role of the mother serves the placenta. It is an unstable hierarchical relationship. It then has to be produced, that is, extracted. Oxytocin serves this purpose. The process of birth is an entire story of self-organization manifest.

As OS puts it, the birth process is a *child production process* due to size effects, and in this particular sense, a series of steps that organisms use to bypass size constraints. It is not reproduction. The notion of reproduction only mildly makes sense when one takes a gene-centred view. And even then, the fidelity of gene duplication and reconstruction is not perfect. What happens is production, which allows for the formation of larger organisms, after the first constraint of limited uterine cavity is surmounted. The child can then grow, unrestrained, up until it reaches adulthood.

From systems thinking, the most optimal option to be taken when a system such as an organism approaches its limit is to effectively remove the limit to allow for more growth. Pushing the pedal harder against an unyielding limit only accelerates system collapse. This is what Meadows refers to when she says that many people need to understand the concept of limits. Growth exposes limits. During the birthing process, the limit presented by the uterine cavity is effectively circumvented by the birth process. The infant organism can continue growing with a limit that is much further than just 9 months of gestation. The mother can also continue to exist, but now, increase in size by virtue of the merger she has made with the growing baby. She had to be sensitive to the size effect of the growing foetus for the birthing process to take place effectively.

An acute sensitivity to size effects is also necessary when we consider the immune system. I had earlier mentioned how the digestive system is part of the immune system. What I left out was how size validates this assertion. The large size of any food substance that is imbibed by any living organism has to be broken down. It is not just broken down because it is a source of nutrients but also because it is a large organism. The large size stimulates the release of digestive juices whose primary function is destroying the food, that is, the organism. The human being's digestive system has to be sensitive to these large organisms up to a point where the digested size does not elicit any such response. Such a size is effectively achieved by mechanical and chemical breakdown processes of digestion. The small size can then permeate the epithelial borders and get absorbed into the body. These nutrients being absorbed are small enough to escape detection by the size-sensitive immune or, if you will, digestive system. The histology and cellular organization of the digestive system are structured in a way that upholds this role in immunity. Nevertheless, not all attempts at destruction will be successful. The unsuccessful ones will either be released by vomiting or released together with excrement.

While a large organism is extremely sensitive to another large organism, it will be significantly insensitive to small organisms. Insensitivity to small sizes allows for particulate exchange at the terminal units such as the capillaries and terminal bronchioles. Their large numbers and, hence, the cumulative surface area of these units effectively maximise the exchange of particulate material. Particles and compounds then merge through particulate symbiogenesis. This creates diversity by introducing new particles and, hence, build on resilience, which is acceptable through self-organization. In a physiologically growing organism, this process results in an increase in size. In organisms that have attained maturity, the practice is an exercise in the maintenance of one's organization – being organizationally closed by thermodynamically open. The growth process from childhood to adult hood shows that living organisms are strictly speaking, not really organizationally closed, since they are gamblers. They would risk merging with a particle at the expense of postponing its annihilation. However, since they possess fairly consistent processes, they can be regarded as organizationally closed biochemically. All these processes generate a statistically unique organism and the benefits all the S's bring. Size, yet again, shows how all the S's are inextricably linked.

This principle of growth exposing a system to its limits can be traced down from the small animals to the larger ones with respect to reproduction rate or gestational period. Organismal Selection would predict that the smallest organisms reproduce fastest and more frequently than the larger organisms. This would be our deduced prediction since the size constraint for a smaller organism is approached much faster than a larger organism. As it turns out, bacteria, which are the smallest living organisms, can reproduce very fast.[32] It would explain why they can infect an individual and have them admitted to the hospital within hours to days. Baker's yeast can duplicate itself through budding every 90 minutes.[33] Among mammals, the gestation of a shrew is around 20-25 days similar to that of a mouse.[34, 35] These two mammals are of near similar sizes. The gestation of a cat is about twice or thrice that of a shrew or mouse.[36] For a cow, it is around 9-10 months, for a horse, about 11-12 months and for an elephant, almost twice that, at around 18-22 months. [37, 38, 39, 40] The prediction gives a coarse-grained outcome that agrees with the data. The larger you are, the longer it takes to get to the limit posed by the uterus, and hence the longer the gestation.

I have to stress that gestation is a marker for how long it takes an offspring to grow to a size that forces the offspring-bearing organism to push it out. I, however, do not hint at the litter number. There is a variation in the litter numbers across various species that I have not had the chance to perform an in-depth investigation on yet. But I would wager that size has a role in it, much like twinning. In this case, gestation is determined by the size of the organisms inside the uterus. So, the shrew might have a short gestation but it does not rival another particular mammal that is almost 100 – 1000 times its size. The Virginia opossum has a record gestation of around 13 days.[41] It also has an impressive ovulatory rate of 60 eggs per cycle. As a result, it can give birth to a litter of 15-20 offspring within two weeks. From the perspective of size exposing limits, the explanation is easily deducible. After fertilization the large numbers occupying the uterine cavity quickly occupy the uterine space and within a short time, approaches the limit. This much is seen when implantation happens for a short 3-5 days before the birth process among the members of this species. To have an idea of how quickly the size limit is reached, it takes a human zygote a week to implant but 9 months before the birthing process. Size constraints do play a role in parturition.

The same explanation is likely to explain the equally short gestation of 15 days by the gray short-tailed opossum, whose weight is 30 times less than the Virginia opossum. What distinguishes the gray short-tailed opossum from the other opossums is that it does not have a pouch.

However, since it hails from a family of pouched mammals, they are known for giving birth to premature babies. All birthed baby opossums are then inserted in the mother's pouch, or in the case of the gray short-tailed opossum, attached to the nipple for about 3-4 weeks.[42] This adjusts the gestation to roughly 6-7 weeks. For the Virginia opossum, the babies stay inside the mother's pouch for about two months, hence having an adjusted gestation of around 10-11 weeks.[43, 44] Our assertion still stands: smaller animals have short gestations. The gestations are short based on the litter number and the limiting space of the uterus and the pouch. There is, however, another peculiarity about the blue whale.

Ranging 10-12 months, the blue whale's gestation is several months less than the elephant. It struck me as odd how the largest mammal should have a gestation period that was shorter than the elephants. On checking the other whales from a simple web search, I was surprised that they also had similar ranges; sei whale at 11 months, humpback whales at 11 months, grey whale at 12 months, sperm whale at 16 months, common minke whale at 10 months and North Atlantic right whale at 12 months.[45] The blue whale is the largest of the bunch but it does not match our predicted trend of gestation periods. The solution came from a logarithmic scale that includes all the whales in the same size category. The logarithmic scale further classifies the gestations of the mouse, shrew and rat at 25-30 days; the lion (110 days) and the tiger (90-112 days); the horse, donkey, zebra and cattle (10-12 months). It would, therefore, make sense for all these whales to have similar gestations. But why would the blue whale have a gestation period shorter than an elephant?[46] I later surmised, as in the case of the Virginia opossum, that the answer could lie in the rate at which this limit was arrived at.

Blue whales are arguably one of the fastest-growing animals for their size. They increase by several billion times from the time of conception to weaning, which takes around 18 months. During this time the calf feeds almost exclusively from the mother, gaining around 200 – 250 pounds (90-113kgs) in a day![47, 48] Comparatively, with regard to maturity, the elephant's growth rate is a tad slower, where it takes one an average of 20 years to attain reproductive maturity. For males, it might be longer – up to 25 years.[49] Maturity for the blue whale is much shorter, that is, 5-15 years. It then becomes apparent which of the two developing foetus approach their limit faster – the blue whale. This is the one reason that could explain the discrepancy which still conforms to our assertion about size. That is, the tendency of an unborn offspring to approach a limit is dependent on its rate of growth *in utero* and capacity of the uterus. For the pouched mammals, it depends on the size of both the uterus and the pouch. Furthermore, it cracks open the idea that the elephant could have found a way to tame this urge to increase in size.

In the interest of extending this argument, I might have used reproduction rate rather than gestation when referring to invertebrates such as bacteria and yeast. Reproduction rate, or duplication rate was the more preferred term referring to how long it would take a bacterium or a yeast cell to duplicate. Bacteria and yeast duplicate asexually. However, we should remember that an organism is the single most unique thing of its kind (just like any other? How ironic). Once we also factor in the tendency to increase in size the idea of reproduction dissolves and what we are left with is production. This should take us back to the definition of an organism or, as we have loved to call them, *living* organisms.

The gray short-tailed opossum (top) and the Virginia opossum (bottom).

Definition of an Organism

Defining an organism as one with the ability to reproduce is short-sighted. Charles Darwin had seen that living organisms had this unique ability. Those that reproduce fastest would then stand a chance of winning the evolutionary lottery. I have a slightly different perspective. The evolutionary lottery is won by striving to avoid annihilation. Success is then marked by the time a new species steps into the scene up until that time when the very last individual of that same species goes extinct. The celebrated Coelacanth, considered a living fossil, can therefore be considered highly successful. Another way of increasing an organism's chances of success is through *production*, rather than *reproduction* as we have always known it. The tendency to increase in size exposes an organism to its limits, constraints that push it to produce. It would then mean that the Gaia can be considered to be an organism, since the various stable ecosystems it possesses, what OS calls organisms and what Lovelock consider organs, have not grown to a size that they become unbearable. Gaia's limits have ensured the symbiotic co-existence between and among the various organisms it harbours. For the short periods that some species have occupied a certain habitat, their continual growth could have exposed the species to its limits, limiting their capability of producing some more. When a system grows and approaches its limit, it either collapses, oscillates into a dynamic balance or circumvents the limit and continues to grow. Let us explore these three fates.

Collapse is a given. It occurs when an organism increases in size but at a growth rate that cannot sustain itself. It is sparked by an imbalance of the interacting systems. Jared Diamond produced another powerful synthesis explaining in broad sweeps, factors that expedite the collapse of societies.[50] On the other hand, a dynamic balance can last for very long, as ecosystems moulded of millions of years have shown us. Among individuals, a dynamic balance is seen when the organism attains maturity. At this point, the rate of growth matches the maintenance. This is the dynamic balance an organism remains in until old age or its eventual death. The third is circumventing the limits set by increasing in size. Through reproduction, the secondary means of increasing in size, what I have dubbed production, it eliminates this limit. Eusocial organisms are the good examples of organisms that have managed to establish efficient ways of increasing their size by this means.

By doing away with *reproduction* and focusing on *production* we dissolve the argument cast against the Earth as one that does not reproduce and hence cannot be relevant in evolution. This is the argument that Dawkins states against Lovelock's Gaia hypothesis. It could appear that the Earth consumes its wastes – but 'waste' is a name biologists use simply because the matter posed a threat the organism. The increasing size of the 'waste' approached the limit for the organism to comfortably contain. It has to be released. Waste is released through the organism's sensitivity to size effects. Earth, too, is sensitive to size effects. Gaia has comfortably existed within its limits for millions of years. Whenever Gaia's contents exceed its limits, Gaia's sensitivity pushes for regulation. The Gaia is much more adept at regulation than humans or arguably any other organism for that matter. An emphasis on production, with reference to sensitivity to size effects, eliminates the need to disregard the Gaia as an organism. Humans have travelled to every 'corner' of this spheroid-shaped organism and they are fast approaching Earth's limit. The Club of Rome has showed how collapse is imminent if the growth of this human organism is not checked.[51] The logic of OS explains why this is a likely eventuality in the future if we become oblivious of our impact on our host.

The game is the infinite type, not the finite one. A finite game is where one plays to win in a one-off play. By contrast, the infinite game is where one plays to continue staying in the game. An example of a finite game is the one-off prisoner's dilemma, while an infinite game is the iterative prisoner's dilemma. By considering an organism as one whose evolutionary role is just reproduction makes their game more of the finite type. Even worse is for those that are infertile, castrated or endangered. What game are they supposed to play if they have not been picked at all? However, in an infinite game, the organism still tries to avoid annihilation long after they have stopped 'reproducing' or even after being declared infertile. Even Gaia plays an infinite game. Production is one of the ways in which it plays this game but this does not mean that organisms should pivotally be described by this one facet. There is more to avoiding annihilation than production (or reproduction, if you're still into it).

Defining an organism based on its ability to reproduce is also myopic in the sense that it downplays the concept of hierarchies. It is the hierarchies that dictate evolution. An organism's goal is to avoid annihilation and this is in line with that of other organisms at the sub-system levels. They are, in short, tied together through hierarchies. If the goals are de-coupled, system failure is imminent. Decoupling is often seen due to size effects. Other reasons could be seen in the death of an organism, which might appear different but upon investigation, is also related to size effects. Consider that when one dies the brain cannot regulate the influx or efflux of materials in the body. It loses its sensitivity to size effects. While the line goes flat, organs have a window period where they can still function – the window period for salvaging organs for donation. The hierarchies' concept allows us to consider organisms connected in this sense. While the larger organism is all but annihilated, the components still struggle to maintain their existence.

So far, I have partly explained how the systems and sub-subsystems decouple and hence break the hierarchies. The body can contain some level of waste, cancer, growing foetus or even poison but past a certain size threshold, it poses a threat. Oxygen is an example of a poison that if an organism is not particularly sensitive to, can cause untold damage. This would take us back to the Gaia, which has its own unique way of dealing with these size-related threats, different from that of biological organisms. We can call the Gaia a geological or even a celestial organism. It initiates ways of eliminating factors that push it to its limit. For instance, it can increase the sea levels so that living beings that stay close to the coastlines of seas and oceans are submerged and drowned. It would increase the temperature by just 2 degrees so that there is a 20-30% death rate. It would then increase the number of storms and hurricanes to dissipate excess heat due to global warming into the atmosphere. The increased temperature by the said 2 degrees centigrade will further increase the production of dimethylsulphonio propionate, or DMS in short, to create clouds and then screen more sunlight to minimise heat entry into the Gaia.[52] In addition, it will push the clouds fast to their limit, resulting in production, which manifests as rain. Increased production of DMS will further influence the winds and ocean currents to shift from region to region creating a cool or warm gradient depending on the region. Simultaneously, the winds would transport nutrients to continents and seas alike. Reproduction as biology defines it, narrows a natural phenomenon that is evident in all of nature, namely, production. The bacterium has its own way of dealing with size effects much the same way the human being but the Gaia has its own unique ways. The principle, however, remains the same – size exposes limits, which would either lead to a collapse of the system; oscillation into a dynamic balance or production.

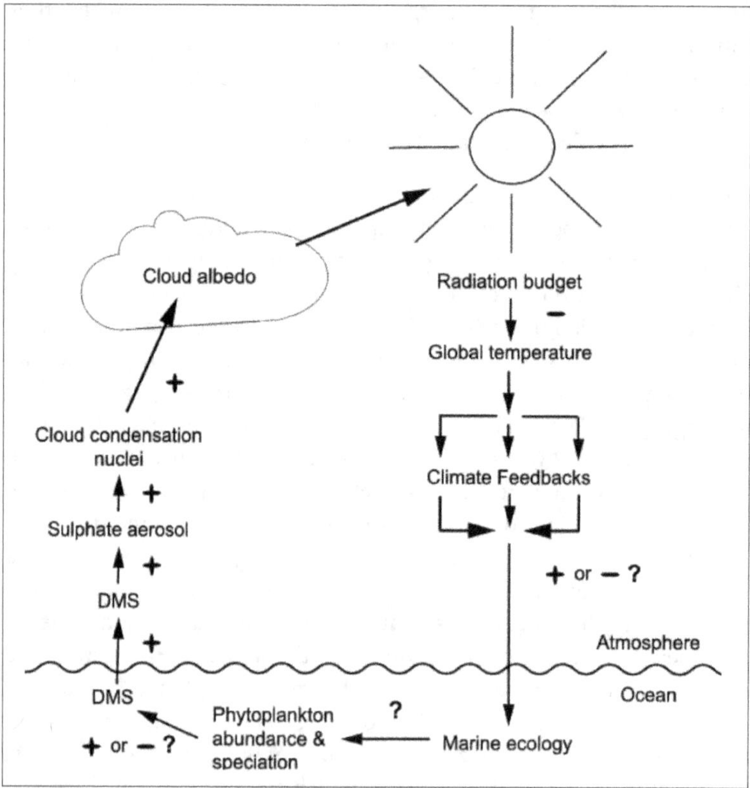

An illustration of how the Gaia produces clouds which end up producing rain in the hydrogen cycle. The sulphate nuclei, generated by the sea phytoplankton, aid in the formation of clouds. The + is the reinforcing feedback while the − is the negative feedback. Similarly, through the cloud albedo effect, Gaia's temperature gets regulated.

This perspective of different means of dealing with the effects of increasing size can be seen from the idea of entropy and systems far from entropy. Broadly speaking, entropy is a state of thermodynamic *equilibrium*. Systems far from entropy are said to be systems in thermodynamic *disequilibria*. When systems are far from equilibria they have a high degree of order, and a lot more information can be derived from them. Hence, the further the system is from equilibrium, the smaller the entropy, the higher the order, the more the information derived. This is in keeping with the criteria used to define living creatures, which are – self-bound, self-generating and self-perpetuating. Fritjof Kapra reiterates that:

> To be self-bounded means that the system's extension is determined by a boundary that is an integral part of the network. To be self-generating means that all components, including those of the boundary, are produced by processes within the network. To be self-perpetuating means that the production processes continue over time, so that all components are continually replaced by the system's processes of transformation.53

What such criteria tell us is in keeping with the deductions from systems far from equilibria. At first sight, the criteria exclude the possibility of considering particles as living organisms. The reason stands that they are not states as ordered as those of living creatures. They have not attained the state we call disequilibria, a state of order and information complex enough to meet these three-pronged criteria. But OS thinks that particles are organisms by their mere existence and that this does not eliminate the idea of living systems as described by conventional biology. Levels of disequilibria, which living organisms display such as the presence of a form of metabolism and genetic code, give them a distinct criterion. These criteria are set simply due to the levels of organization. A cell will be different from a tissue, a tissue different from an organ, an organ different from an organ system, an organ system different from a living organism (biologically), a living organism different from an ecosystem and an ecosystem different from Gaia. These are different levels of disequilibria with different degrees of order and hence different degrees of information.

Different degrees of order separate the invertebrates from the vertebrates, prokaryotes from eukaryotes, warm-blooded animals from cold-blooded ones, mammals from birds and reptiles from amphibians and fishes. A criterion such as that of a genetic code *and* metabolism or a three-pronged one described by Kapra, is broad enough to include all the living organisms that we know. However, it could also blind. For starters, one becomes blind to the possibilities of self-similar properties between the 'living' and 'non-living'. As a result, the quark would be different from the electron, and these two different from an atom, the atom different from a molecule and a molecule different from a complex compound. They appear different due to the different levels of organization. They are also unique, as OS asserts. They however display a *fractal behaviour of nature*, much like the *fractal geometry of nature*, where they all display tendencies of avoiding annihilation. What's more is that from the self-similar perspective, OS reveals that the three-pronged criterion cuts down to the small particles that are considered non-living and extends past the boundaries of the Gaia.

Before we explore each of the three criteria, here is a brief reminder of the five invariant properties. The first is the most robust – the map. I have shown how the map remains invariant, regardless of the size or type of the organism. Secondly, the tendency to avoid annihilation, is invariant, regardless of the size on all scales. The four-fold pattern of preferences, for instance, shows striking similarities of how atoms behave much like organisms. Thirdly, the definition of an organism remains the same. Fourthly, the ability to detect the presence of an organism, by introducing a credible form of a threat, is also invariable. The final one is that the inherent tendency towards increasing size exposes the limits of the organism. We can, however, try go along with the criteria that Kapra has described to see if a particle such as an atom or massive bodies such as the Gaia satisfies the three criteria of self-boundedness, self-generation and self-perpetuation.

The atom is self-bounded, selectively allowing electrons to go in or leave. The cell is self-bounded, selectively allowing which particles to go in or out. The earth's atmosphere is self-bounded, allowing various UV rays to go in or out. This boundary is an integral part of the network that forms all these entities. The first one checks out.

A superficial analysis might hint that the atom is not self-generating like the cell or the Gaia. The cell undergoes metabolism to renew its boundary and the Gaia has its geological metabolism in various cycles resulting in the generation of the atmospheric components and its boundary. Nevertheless, the atom has self-generating processes. The mobile nature of the electrons in relation to the proton constitutes the atomic 'metabolic' processes that allow it to maintain its structure and its boundary. The atomic components, such as the nucleus or electron do not need generation like that of the cell or the Gaia. But their interaction is needed for the atom to maintain its structure. Consider an atom with immobile electrons. It ceases to become an atom. What if we invert it and have a nucleus and a constantly moving electron? Through the constant movement of the electron in relation to the nucleus, the atom indeed becomes self-generating. The atom, the cell and the Gaia have to be thermodynamically open for their respective metabolic processes to take place. These processes then maintain the organizationally closed structures and the boundaries. The second criterion also checks out.

Regarding self-perpetuation, the production process as biology puts it, is always on-going so that all the contents are continuously replaced over time. The Gaia does this too. The atom does not need to continually replace its components, even though it can, by temporarily switching from a charged particle to a neutral one by releasing or gaining electrons. It does, however, need to maintain its processes. It does so by being thermodynamically open, just like the cell and the Gaia. This crosses out the last criteria.

Thus, even the three-piece criteria are satisfied by particles after deeper analysis. The perspective of levels of order at different scales conforming to different levels of information is needed to arrive at this conclusion. Whatever criterion suits you is a matter of choice or convention. Defining an organism as OS does practically includes all of them. Thus, it does not eliminate their properties as the sciences have always known, but upholds the rich diversity that OS proclaims – the fundamental unit of diversity is at the level of the organism, regardless of the scale.

How do we learn?

I find the idea of an organizationally closed pattern and a thermody-namically open system constitutes a framework used to identify all organisms. Einstein had already shown how matter can be converted to energy ($E=mc^2$). Organisms can therefore take in particles as they so deem fit since this itself is an exercise in thermodynamic openness. Simultaneously, it allows for the maintenance (although imperfectly) of an organizationally closed pattern. I have swum with this idea to this point but now OS introduces a certain twist to this framework.

A new organism is formed every time a new merger is established. Organizationally and thus formally, it remains closed, but empirically, it is different. Consider the structure of water comprising two hydrogens and one oxygen. Writing down a chemical equation hardly captures how these particles constantly interact as water. They form and break bonds between them in a continuous aqueous state, but these shifts are not strong enough to alter the constitution of water. The chemical for-mula shows an organizationally closed molecule, but empirically, the water is ever twitching, and particles constantly shifting. The idea of be-ing organizationally closed is a great approximation in appreciating the unchanging state of compounds but it hardly matches reality. It further contrasts with the idea of self-organization, where particles are continu-ously facing universal perturbations. In reality, particles are ever on the move and interacting with other particles. If the constitution is strong enough, these interactions hardly make significant changes appreciable to the naked eye. What this means is that every time a human being in-teracts with oxygen through respiration, it forms a different organism, even though the organization remains fairly intact. But the question then becomes, how does a multicellular organism such as you or myself with multiple levels of interactions with particles remain self-referring?

Recall that particles can be so small that they are hardly recognised by the highly sensitive immune and digestive systems. These systems are sensitive to size-effects. Atomic or molecular mergers are small changes in comparison to the overall size of a multicellular organism. I will still be able to recognize my mother even after she applies the latest skin lotion that my brother bought her. Secondly, there are cells that are permanent in human beings. Only on special conditions do these cells replicate or regenerate but until then, they are generally 'settled in their position'. These are heart cells, skeletal muscle cells and cells of the nervous system.[54] By permanent it means that once they are damaged, they get no replacement. They are different from other cells such as epithelial cells that get replaced every other time. Permanent cells help in understanding why organisms can be self-referring, and here's how.

There are permanent cells and renewable cells. We can think of them as two entities in a hierarchical relationship. The purpose of the upper hierarchy (permanent cells) is to serve the lower ones (renewable cells). The brain thus does an audit of what these renewables will need for their short stay and in the process, postpone their annihilation. At the same time, the purpose of the lower hierarchy serves the upper one. The renewables are on the front-line, receiving jabs from the universe. For every bout survived or failed, it is marked as a learning experience that the permanent cells pick up. The gradual changes amount to the learning process of the overall organism, the self-referring organism. Thus, the self is in constant flux, learning constantly, with every entry and exit of particulate organisms. This is the only plausible way I can conceive that organisms learn, by constant interaction with other organisms (particles). This would mean that an organism is constantly constructing its understanding of its own universe. By having a permanent set of entities such as the DNA or the neurons, the renewable entities can be utilised in a way that an organism can test its tentative theories and adjust its world by error elimination, thereby constituting a learning process for the self-referring organism. At the very core of organisms there must be a fairly permanent entity. As it increases in size, it may decide to forge mergers with other entities temporarily and hence postpone the annihilation of all the involved parties. These mergers, therefore, are brief classroom sessions where organisms get to learn more about their universe.

Over time, some interactions become fairly redundant. They acquire a similarity to the first encounter, so there is little to learn. Take the example of a coin flip. If you do it once and it lands on heads and a second time, it again lands on heads, it can appear rigged. But if you do it 100 times, it gravitates towards the formal outcome, where heads might have appeared 54 times and tails 46 times. Switching the toss to 1000 tosses will hardly yield a lot of learning. The repeated instances, over time, do not account for significant learning – they appear redundant. But the first instance is overwhelmingly new. During these interactions, learning is significant. For instance, when a baby is born it has not been breathing throughout its stay in the womb. Once it gets out, the interaction with the oxygen and carbon dioxide particles outside the womb translates to a significant learning experience. It then shifts its red-cell haemoglobin concentrations to match this new experience. Over time breathing becomes a habit and redundant, with little learning. It later learns how to pass stool when the size becomes too much inside their bowels, or it goes into a fitful cry until an experienced mother or doctor improvises a suppository. The initial experiences are significant moments of learning that when repeated enough times, become reasonably redundant.

At this point I would want to make a wild, wild speculation but bear with me for a moment. What if every particle has a form of 'memory' that it has stored from its various experiences? What if this memory explains why chemical reactions are so regular rather than novel? Moving from lower to higher hierarchies, particles existed long before the DNAs or RNAs. The very first particles that were able to form the DNA or the RNA experienced this type of reaction and were able to do it again. This would ensure the fidelity process of DNA replication and regular cell cycle processes. Each primordial particle is more permanent than a DNA sequence. They therefore could have the potential to have much more preserved experience, such as having consistent outcomes in chemical reactions, than the DNA or RNA. The DNA or RNA would then have more preserved though richer experience than the particle, all the way to the living organism. What if this process initially started outside of Earth and was later, through a series of redundant steps, shared by the particle constituents of Earth to the point of emergence of 'living' beings? This would explain the self-organization of viruses latching onto bacterial or eukaryotic genomes with high precision to facilitate viral replication. It would explain the consistency of outcomes for every sequence we make or chemical reaction we conduct in the labs. The particles could have learnt this early enough and resorted to the outcomes that we now observe due to the experiences they had. It would explain why memories can be stored or literally remembered, while at the same time how new ones can be constructed. DNA or RNA could not be the only molecule that archives information.

What if this is what particles have always been doing? If we begin to think of particles as organisms, then it just might be the case. Coupled with the conversion from *reproduction* to *production*, it also gives an alternative perspective about the idea of the *unit of selection* (the gene) and the target of selection (the organism), since every particle is always targeted by the universe. I will; however, not stretch the possibilities past this point.

My slight detour in speculation should remind you of the role of the immune system, which was to have a fairly good understanding of self, especially during periods of good health. This role ensures that the pattern of the organism is maintained or gets slightly modified for the better with every interaction with other organisms. With these small bits of learning, self-identity remains. But since there is a constant interaction with particles, *physically*, the idea of self-perpetuation becomes a moot concept. It only begins to gain validity when referring to an *organizational pattern*. Having an oxygen atom replacing another so that the formula for water remains H_2O means that the pattern has been preserved even though a particle has been replaced. So long as an organism has merged with another, there is the formation of a new organism. If the interaction is similar the shift from the old to the new organism is negligible, hence the ease in self-identification in spite of the numerous similar interactions with particles.

Cornelia Bargmann, a highly distinguished geneticist whose work I admire, has shown how little bits of data can elicit similar patterns in her favourite model worm, *C. elegans*.[55] Even when the worm was exposed to small bits of a certain compound, there was a neuronal pattern similar to that produced when it was exposed to the complete compound. The experiments have shown how there is negligible learning even from a fragment of something that is representative of a whole substance. For instance, exposure to half a slice of bread would elicit responses similar to a full slice of bread. It implies how patterns are more relevant in learning through interactions with particles. The patterns are analogous to the tentative theories that organisms formulate before testing them out in their universe to see if they are valid or not.

This idea underlies the broad detection and elimination of pathogens by the immune system. Once an immune cell is exposed to one strain of a bacterium, it can formulate a strategy of eliminating it that will strike the current and other strains that have a similar 'appearance'. Should there come a time when a slightly different strain attacks, it would then kick start its response with the previously preserved strategy to test if it is as efficient in eliminating the new bacterium. The preservation of strategies is evident through a special type of immune cell known as the memory T cell.[56] The same learning strategy further explains why Bargmann asserts that the sensory system is one of the most rapidly changing systems, since it is the one that gets to interact the most with its environment. In short, we get to learn a lot through our sensory system. Coupled with pattern formation and memory, which equates to the tentative theory and error elimination stages of the Popperian schema, we can learn a lot about ourselves and our surrounding universe.

Since it was Fritjof Kapra who introduced me to concepts of non-linearity and the connections evident in life, I have to admit that he was the one who clarified for me that it was networks that were more if not equally as fundamental as particles. Patterns highlight systems perspectives more than a narrow focus on particles alone. The networks created by the particles resulted in organizations that Francisco Varela and Humberto Maturana formulated in their ingenious idea of autopoiesis. In the same spirit, I have to insist that the map I have constantly ranted into a monotonous drone, remains unchanged in spite of the uniqueness of every particle (s). The explanatory power of OS has scaled past disciplines that are thought to be very distinct from each other. In fact, I even think that it scales past the need to distinguish Natural Selection as different from Sexual Selection.

Sexual Selection

Charles Darwin felt that Sexual Selection was different from Natural Selection, so much that he considered them as separate. In Sexual Selection, an organism develops particular preferences for certain traits in the opposite sex. In this type of selection, the organism decides who to mate with, unlike Natural Selection, where it does not factor in the fact that an organism has preferences. Like a sculptor, Natural Selection chips off David from the stone it is given. David does not crack the stone and walk out of his concrete egg.

Darwin cites a special example from monkeys that had always intrigued him. Male monkeys changed colouration during breeding season, which he related to the fanning-out of the brightly coloured peacock feathers.[57] If a peacock has more 'eyes' in its feathers once they have fanned out, the peahen would prefer it to another with fewer 'eyes'. Such traits, over time, end up lacking the adaptive effectiveness seen in organisms. For instance, the peacock does not necessarily need the large expanse of a tail with much adornment. However, according to the single facet of Natural Selection, which is reproduction, the many 'eyes' grant it access to many peahens and thus indispensable in Natural Selection. On the other hand, Organismal Selection does not need to be split into a separate version to describe sexual behaviour. It can still adequately explain the sexual behaviour of animals. For instance, in OS, the organism creates its own path and is actively involved in shaping its outcome. David eventually steps out of his concrete egg-shell, but after millions and billions of years of several organisms working their way in avoiding annihilation. It is important to note that it is widely accepted by many scholars that Natural Selection is creative (chipping off bits from a stone to curve David). We are made to believe that this sculptor is blind but makes impressive art. On the other hand, OS asserts that the creative force is a cumulative, intentional and random effort by the different organisms to avoiding annihilation (the cumulative success of avoiding the hammer and the chisel). Sexual Selection is a powerful explanation of what OS asserts.

The peacock and its massive tail of 'eyes'.

Before we dive into the details of Sexual Selection, there are some issues I would like to clarify about it. In the interest of clarity, the decision to use the word *reproduction* is a replacement of *production*. Hence, anytime I use the word reproduction, what I mean is production, as I have previously described. Now, the first issue has to do with the fact that organisms have to interbreed and give birth to viable offspring. They ought to have the capability to engage in sexual intercourse for there to even be Sexual Selection in the first place. Not all living organisms undergo sexual reproduction. Bacteria and fungi, for instance, experience asexual reproduction. The other issue is that of parthenogenesis – the ability to produce viable offspring without fertilization from the opposite sex.[58] It is an asexual behaviour seen in several eukaryotes such as some insects, plants and even Komodo dragons.[59, 60, 61] Patterns where these species engage in this form of asexual behaviour is seen in extreme climatic conditions or in situations where the other sex is not available. This pattern of behaviour can be explained by OS as the tendency to increase in size, even in the absence of the male counterpart. Female Komodo dragons can reproduce in the absence of the male counterparts. This ability strikes at heart of self-organization and even more, the tendency to increase in size even in the absence of males. These two processes, asexual reproduction in the microbiota and parthenogenesis evident in some plant and reptilian species, could arguably be precluded from the Sexual Selection theory.[62] Rather than sweep this phenomenon under the rug, I argue that the question that should be asked is why such behaviours are present to date. The answer, I believe, lies partly in the tendency for an organism to increase in size by surmounting the limit that physiological growth brings. The other limit they surmount has to do with the dependency on males for fertilisation, which bars very many species from increasing in size. These two reasons back our thesis that organisms strive to increase in size.

Among those that can sexually interbreed, the Sexual Selection explains a lot even though Darwin noted that it often resulted in maladaptive features. Ronald Fisher, in his runaway model, further clarified Darwin's insight.[63] For instance, the tail of the peacock would variably increase in size every generation regardless of its harboured risks. One crucial mishap that a long tail brings with it is that it hampers the adept stability and flight competence of a bird. Why this risk is taken by the peacock can be explained by Sexual Selection, which would grant the peacock a reproductive advantage. It would attract many peahens and, according to Natural Selection, reproduce more compared to the peacocks that had smaller tails. What had always bothered me was why many of these maladaptations were related to an increase in size. During the breeding season, the male frogs gather at the edge of a pond and start croaking. The females would then pick the frogs with the deepest croaks. A deep voice, Steven Pinker shows, is significantly correlated with large size. Voice then can be used as a surrogate marker for size. Some studies have however been able to show that there are special conditions where the deep croaks are forsaken for the fast ones.[64] For the peacock, it is the *large* fanned-out tail. For the antelopes, it was the *large* protruding antlers. For the long-tailed widow, it was the *long* tail. For the lions, it was the *large* mane. For the high school and campus students, it was that giant of a guy that majored in sports.

Fisher developed a wonderful principle that explained this phenomenon. He called it the runaway principle and this is how it works. The female would desire a trait that it cannot manifest and the male would portray this trait. It would then reproduce with this group of individuals. In the next generation, the same male individuals would be present while at the same time, the females who displayed this desired trait that they cannot manifest would also exist in greater numbers. The runaway is the continuous reproduction of females who would only select for this extreme feature, while the males with the smallest or average manifestation of the trait end up left behind – they get little to no group of females to breed with. The self-reinforcing momentum of both of these sexes complementing each other can lead to a runaway event of organisms with maladaptive traits existing at the expense of those with adaptive traits. The peacocks would have long tails, the lions larger manes and antelopes large antlers just so they can get laid. It is a well-established principle that has been accepted by a wide array of evolutionists.

Enter Amotz Zahavi who coined another principle, the handicap principle, which also bears relation to size.[65] A large-tailed peacock would signal to the opposite gender that they can bear this maladaptation as a handicap and still survive in the wild. It was a clear albeit costly signal that they are strong enough; a bearable size and a bearable strain. This principle again, coincides with the examples listed to explain the same phenomena as the Fisherian runaway principle. However, in my analysis, it appears that the handicap principle has more explanations other than just size and thus proves superior to the runaway model. Bower birds are an example. They can set the mood for breeding by decorating their houses in various adornments.[66] The nests they make are a thing of beauty and also a signal that some Bower birds are large enough to carry various material over long distances, create them neatly and defend them from creatures. This is a signal that comes with size – you cannot ward off predators or competing birds if you are of a smaller size. But it also shows that these features can be displayed outside the animal, as in the case of the bower bird nests. It displays that the cost it takes to make the nest is bearable, signalling to the females that they are reliable mates. It therefore follows that the houses are expected to become bigger or prettier and territories larger with every generation, under certain set limits. The territoriality extends even to human beings. In African traditional societies your respect was matched by your property. Back then, children and wives were considered property of the man. The more he had, the more it rang clear as a signal that you can bear with the size.[67] Size sends the signal that one can live with the demands that it brings.

Thus far, it has been widely documented that these two famous principles can adequately explain Sexual Selection. Same manifestation, different interpretations. Organismal Selection offers a third. Size, according to OS, brings with it two outcomes – increased efficiency and delayed pace of life or if you will, prolonged existence. For every organism there is an inherent tendency to increase in size. There are benefits to increasing in size besides getting laid. From the works of West, having a large size organ would imply efficiency of energy and from relativity, the prolongation of existence. It, therefore, can explain the attraction to larger manes, the long tails and the extended, convoluted antlers. It also explains how and why they can be manipulated even among human beings. The larger the house one has, the more the opposite sex partners are likely to get attracted to the individual (females are more likely to be attracted to the males in such a scenario). The role of the large house serves the man in having more ladies as sexual partners, and the role of the man serves the house by preserving it or even adorning it.

This is a stable hierarchical relationship, constituting a merger. The reverse would largely be seen when young men seek rewards from older, wealthier women. Size, in terms of wealth, implies efficiency of resource use. Larger organisms display an efficiency that is lacking in smaller organisms. Taking cities or even companies as organisms, it explains why the large ones attract so many people compared to towns or villages or small businesses. Size implies efficiency. It also implies the ability to push harder or longer in spite of the size constraint. Ergo, handicap. If we are to consider the runaway principle, the role of the females serves the males with the particular trait and vice versa. This constitutes another stable hierarchy. The role of the city or company serves the stream of immigrants or applicants respectively, and vice versa. Another stable hierarchical relationship. The unique aspect that OS introduces is that both sexes serve each other in exaggerating a particular trait in the interest of increasing in size.

Although somewhat similar, the interesting distinction between the previous principles and that described by OS is that there are nuanced reasons why size is picked. It is not just simple desire as in the runaway principle or an ability to withstand a handicap as in Zahavi's principle. It is that size increase signals efficiency and tendency to last for a longer period of time compared to other smaller comparatives. Since size is a cosmologically known way of delaying one's doom from the times the observable universe began registering particles, it is a relatively preferred option universally. It is the same inherent tendency to increase in size that pushed the sea creatures to increase in size past the weight-bearing limits set for those that live on land. The benefits of buoyancy gave them the leverage to continue increasing in size – it pays off. Defining evolution as the generational passing of beneficial traits will view an increase in size as a maladaptation. Defining evolution as the ability to avoid annihilation will view an increase in size as an adaptation. Size is a reliable signal.

A subtle but valid difference between OS and the Handicap Principle is the signal interpretation. In the absence of another organism to decipher the signal, size cannot act as one. In such moments, it only serves the individual in delaying its annihilation. Therefore, signals only serve their purpose in the short-term, when there is another individual to interpret it. For example, the more horizontal the black stripes on the rear of the zebra, the more it signals its broad hind-muscles. A lion would be wise to avoid hunting such a zebra. A merger that serves both entities is then temporarily formed – the role of the zebra serves the lion and vice versa. This is a stable hierarchy. The lion would have better chances chasing zebras with more vertical stripes. In the absence of the lion, the zebra with the more horizontal rear stripes only serves itself by delaying its annihilation since it is larger than the zebra that has more vertical rear stripes. Thus, in the long-term and in the absence of another individual, size serves the organism in delaying its annihilation.

While we are in the topic of sexual reproduction, there is a group of creatures that display a rather unique behaviour – cannibalism during or after copulation. Take for example, the praying mantis, which feeds on the male during copulation.[68] Natural Selection has offered various credible reasons for this behaviour the leading one being the provision of ready nutrition. I would argue that no organism with a goal in trying to avoid annihilation would willingly allow itself to get consumed. Therefore, OS offers another reason – that the female anticipates an increase in size and a consequent strain that comes with an increase in size. The increase in size stems from the growing offspring inside her. The strain regards her nutrition and, simultaneously, that of the unborn offspring. The unborn might demand much more than the mother can deliver in the absence of readily available nutrition. Feasting on the male during or after mating is the readily available option that the female resorts to. Such a feast would logically result in an increase in the size of the mother. However, the more pressing threat is that the offspring will be growing much faster than the mother. The mother has to find a way of increasing in size fast! Hunting with such a burden inside you can be tedious, and even then, success is not guaranteed. There is a credible threat to its life. It has to assert its autonomy. Therein lies the other reason for feasting on the male. The female would then eat the nearest available food in order to offset the projected nutrition deficiency. It solves two problems with one solution – the mother extends its stay by increasing in size and contains enough to supply the growing offspring. If there was no ready nutrition both the mother and the child would have died even before the birthing process. Size gives them the nudge to last long enough until the birthing process. It then gives the mother the much-needed solution to prevent its anticipated doom. This is the female's side of the story. What about the male's side?

In the male's lifetime, the odds of getting a female mantis to mate with are very slim. But even slimmer is the ability of the female mantis to survive in the absence of nutrition for itself and hence for the children. The *adept* defector male mantises that escape being eaten do not have children who propel this defector behaviour. Those that survive are the ones that propel the *inept* defecting behaviour, hence, end up being eaten. They provide the female mantis with sufficient nutrition to bear the offspring and possibly enough to last the mother with more nutrients after the offspring are born. The idea of escaping annihilation over generations ended up with the mantises that are so inept at escaping that they very often get eaten during or after copulation. It is an inescapable consequence since this is the only way the male mantis can increase in size after reaching adulthood. It is also a losing gamble since this single means of increasing in size results in death. This cannibalistic behaviour is not only seen in the mantis but at a more lethal dose among spiders.

Spiderlings often feed on their mother just like any unborn offspring does. As for the adults, a good number of species have their females feeding on the males during or after mating to provide this ready source of nourishment just as we have seen among the praying mantis. There is, however, a rare group of spiders that consume their mothers to the point of death in a process known as matriphagy.[69] It is a testament to the autonomy and dependence of the individual organisms. It often happens when the offspring are already hatched. This feeding pattern initially establishes a stable hierarchy where the mother feeds the spiderlings. Later, the hierarchy is 'destabilised' as the mother is eventually feasted upon completely by the offspring. It might appear that the hierarchy established is unstable only with reference to the mother and her young ones. But these spiders are usually social insects. In their small community, there are virgin spiders that cannot give birth and take up the role of bringing up the hatchlings together with the mother. Once the demand for more food increases, the mother offers herself to be eaten by the young ones. By viewing the entire group as an organism, this sacrificing process is hardly any different from autophagy. Autophagy is the process of 'feeding oneself using self', often during periods of food scarcity. From this perspective the hierarchy is somewhat stable – the mother that serves the larger organism (the entire community of spiders) and the community that benefits itself by feeding on itself, giving it a chance at survival. The spiderlings that feed on the mother increase in size faster and thus moult earlier and hunt bigger prey. This increase in size of each of these individual organisms enhances the efficiency of the overall organism (community). Ergo, the much better term production rather than reproduction.

Before we wrap up this topic of what by now we should rename to production and not reproduction, size effects and constrains of size could further explain other peculiarities. The first one I noticed was while rearing chicken at home. I had wanted to start a mini-business of rearing broilers as a side hustle while still in campus. It turned out to be more of an experimental set-up than a business venture. I learnt invaluable lessons in business, but more than that, I became an observer of chicks and chicken behaviour.

To ensure the chicks were warm, I stored them in my bedroom, where I spent most of my time. Initially, I noticed how the chicks could hardly recognize water when I first handed it to them in a bottle top. After one poked at it and later found out that it was water, the rest followed suit. They were well aware of what it was the second time as they scrambled for it even before I laid the bottle top down. I, therefore, tinkered with them a bit. One time I decided to change the lid that they used for drinking and the same thing happened. Only after one of them tried it out and discovered that it also bore water did the rest attempt drinking it. This sparked my rather peculiar interest in chicks and chicken in general with particular interest in evolutionary behaviour.

I had keenly started noticing a behaviour pattern when my mother decided to make the most of the chicken coop I had made in the backyard. She got a hen, it laid eggs and three weeks later, had a group of chicks. The mother was very protective of them. She would spread out its wings and raise its feathers every time anyone approached her or her chicks, ready to strike. When feeding, she would show them the right meals and the chicks would only feed after the mother had confirmed that it was appropriate. They would also respond to a specific shriek from the mother, and immediately run to the mother's feet, in hiding. The mother would then crane its neck out, on the lookout for danger. Only after it had lowered its neck and started clucking would the chicks continue wandering about. The mother and the chicks coordinated behaviours and hence communicated. They were a single organism. What struck me as odd was how over time, when the chicks had grown to a size indicative of the chicken equivalent of the adolescent stage, the mother would turn hostile. She was no longer motherly anymore. It was odd how it was once protective, and another time hostile. I could not understand that. From an evolutionary viewpoint, the divergence from care to hostility seemed not to have an explanation. I later learnt that lions behaved much the same way. When the lions had started showing their manes in their adolescent age the alpha males would chase them away or they would leave to establish their own territories.[70] It was only after I had formulated the idea of size constraints and production did it make sense. The explanation I had in mind bore similarities to that of Dunbar numbers and societies.

The chicken and the chicks were a single organism that could contain itself based on the available resources. The mother would show them which food to take and protect them from any danger. The danger, apparently, was any of my family members who only came out to give them food. However, as the chicks grew and became bigger, the resources started dwindling. The individual organisms felt that there was a threat and decided to assert their existence. Due to the constraints of size, the behaviour shifted from cooperation to competition. The chicks were becoming so big and demanded so much more food that it left little for the mother hen. It was at the point of criticality, at the brink of the organismal size – the organism being the mother and the growing chicks, that production ensued. The mother and the adolescent chicken fell out and turned into rivals for the same food. Often, the adolescent chicks would move as a group, a single organism, maintaining a cautious distance between them and their mother hen. The initial organism had self-organized to a point of criticality, resulting in an avalanche – a fall-out.

The same explanation applied to the alpha male lions, who would chase away the adolescent males from the territory, since they might push the organism's size to the limit, leading to collapse. To eliminate this limit they would be chased away and minimise the fast approach towards the size limit. Resources can then be sufficiently abundant for the remaining pride. The size constraints in these scenarios are an example of a special type of production. It is a form of 'labour pain' where the parents become extremely sensitive to the increasing size of their offspring. It then develops a harshness geared at 'pushing' the growing adolescent chicks or lions. The labour pangs are the regular encounters with reduced meals with every hunt or scavenge. The parent has to 'push' them out for her to avoid annihilation. A keen eye will notice that once more, we have a replication of steps that are likely to have started at the beginning of life – initial cooperation followed later by competition. However, unlike the birth process, the adolescent children or young adults are sent away.

Among humans, the situation is also similar but with certain caveats. Just like the alpha male or the mother hen, size constraints force parents to push their children away from home. This usually starts at that time when parents and their adolescent children disagree. But when the parents get old they would rather have the children, now adults, at home to help them. They also have constraints that they cannot handle at their age. It turns into a tango between early cooperation, mid-competition and later cooperation. This last part only developed with the prolongation of man's lifespan. In all these situations, the idea is the same – anything to avoid annihilation.

The trend has been the genesis of social classes and castes in different human societies.

It has been suggested and variably observed that the rich have fewer children than the poor.[71] The reason I have often heard about such disparate outcomes was that the poor had a lot of time in their hands and would rather copulate while the rich were always busy and had little time for that. After forming my theory, I could not quite agree with this logic. The reason is that the poor would rather have children since by themselves they are pushed to limits that they become sensitive to, demanding a quick means of escape. Getting children is not an added constraint but an investment. It is a way of having more hands bringing more food to the table. Walk along the streets of Nairobi and this is what you will find – a mother seated at a distance with the kid seeking food or money from individuals traversing those streets of by-standers. There are other reasons why the option of getting more children is preferred such as religion or tribal customs, but the explanation I present knows no allegiance to personal belief. It is another nuanced example of how increase in size enhances efficiency; having extra children pushes away the limits, reduces the concerted efforts at getting by and the returns are evident – this could likely explain why more poor people end up having more babies. They are an organism, a single organism that can be detected by threatening any one of them. It is not that they have a lot of time. It is that they have close to none. All of it is consumed trying to get by through whichever means. Meadows said that at any one point the input that is most important in the system is the most limiting. This explains all these ingenious ways of production. Talib Kweli said it best: we do everything and anything just to *Get by*. Getting more children is much like the dance of the bees, the trail of the ants or the colony of termites – the more they are, the easier it becomes to get enough for each other.

What of the rich? They also have developed their own organism – their empires, estates, companies, or whichever property they own. Getting more children should then be the case since they do not have any size constraints. It might appear so, but the roles here are reversed. This is only an interpretation that can be made from a relative basis, that is, from the viewpoint of the poor who see that the rich as the kind of people who have 'everything'. In the real sense, the rich are extremely sensitive to their size just like the poor. They are so sensitive to it that the idea of getting more children seems to suggest that they will not just be costs but and division of their wealth. Children will reduce the size of their wealth. They would not want that. The resources would begin to dwindle. Even more, is that the children are an instant reduction of one's wealth that has taken a long time to develop. For a good number of them, the prospect of getting more children would be pushing them to constraints they are not comfortable with. To avoid such constraints, they will not have many children. Just a small number, and to this small number they would want them to continue what they have built, so that they are continuously converted from an expense to an asset. This interpretation has its caveats, as having many children in the past traditional African communities was taken as a sign of wealth. There are also other wealthy couples that have very many children, an outcome that can be explained away following a better understanding of the very many family traditions and individual beliefs in different parts of the world.

The idea of children as assets or expenses is not a new thing among humans. Before the industrial revolution, if a child was weak the parents would abandon it as it invited more constraints or risks than benefits. It was more like the wild back then before the introduction and execution of the universal bill of rights. The message is clear – *organisms would rather evolve for resilience than production.* If production results in resilience, it will be prioritised. If production destroys resilience, it will not be considered. Resilience is also the theme of the story behind menstruation and menopause in humans.

Overt menstruation is only experienced by genetically female human beings and close relatives such as chimpanzees.[72] Other female mammals have oestrous cycles. In menstruation, the female starts shedding the outer lining of the uterus if there is no fertilization of the realised ovum with every ovulatory cycle. This is a process occasionally accompanied by cramps – painful muscle contractions of the uterus. The mystery behind menstruation plagues very many evolutionists with several hypotheses having been suggested.[72] Of all the explanations I encountered, I was greatly impressed by Zahavi's interpretation regarding the monthly flow as a signal of health, one that cannot be faked by a sickly or old female. Organismal Selection can however offer another alternative solution to the mystery. Firstly, menstruation is an indicator of near-certainty that a female is not pregnant. Its reliability is supported by the fact that it is very regular among most women. Secondly, human beings are unlike other mammals whose ovulation happens the moment copulation beings.[73] This form of induced spontaneous ovulation usually enhances the chances of fertilization upon copulation. In humans, however, ovulation is concealed. A woman can get some vague feeling when it is about to happen, from hours to days. However, there is no knowing *when* it will happen or, in retrospect, the time it actually occurred. Therefore, copulation among humans has a risk of not achieving fertilization. Menstruation can then be a reliable marker for fertilization that the female can use to determine if it is pregnant or not. The second one is a tad speculative. The regular shedding and the occasional painful cramps could be a means of engaging the uterus to enhance its sensitivity to what might happen during size constraints that begin to take shape in during the pregnancy stage and eventually during labour. The muscle needs to remain sensitive to size-related luminal changes as seen in other luminal organs (such as heart chambers, carotid sinus, urinary bladder, various parts of the digestive system and its associated reflexes to name a few). I call it speculative since substantive evidence has shown that there are specific pathways (such as prostaglandin pathways) that cause the contraction of the uterus during the process of birth and are also involved in causing pain during menstruation.[74] Presently, I would not completely consider this as the reason for menstruation, but speculate that it is a possible explanation that might have an evolutionary role.

What is more striking than menstruation is menopause. It is the cessation of menstruation when all the ova have been released in the lifetime of the woman.[75] Menopause, much like menstruation, only happens in humans (and a few other marine species). Why? For a long time, the 'natural' human lived to around 30-40 years. Technological and healthcare advancement have prolonged this lifespan. Menstruation and menopause can therefore be considered evolutionary leaps of self-organization. Menstruation, possibly, for the reasons I have already given, which is, to maintain sensitivity to size-related changes. Menopause for *also* being sensitive to size effects, in the following way.

Having children past a certain age poses threats to both the child and the mother. Even at an early age, say mid-twenties, pregnancy still is a risky state for the mother. You can acquire an otherwise mild infection that can lead to serious complications; blood clots that affect the normal function of the child and the mother; hypertensive states that are otherwise absent and even develop a temporary form of diabetes that increases the chance of the mother acquiring diabetes long after the pregnancy. Eliminating such threats by ensuring that there are no more pregnancies could be what menopause is all about. There will, therefore, be no need for a surrogate marker to check for lack of fertilization. At the same time, there will be no cramps or menstrual flow according to the size-related idea I just explained, since the effort will no longer be needed in having a sensitive uterus. But more than that, it captures the goal of the organism – the need to avoid annihilation. Self-organization equips organisms with the capability of withstanding universal perturbations – menstruation and even more, menopause does just this for adult and elderly females.

The causes for the ever high maternal mortality rate have been summarized into four 'toos' – the women are either *too* young, *too* old, have *too* frequent children with minimal spacing or have *too* many children.[76] Menopause has proven to be evolutionarily stable in combating the global maternity mortality rate by denying mothers past a certain 'old' age from conceiving. If a mother has to cease the flow of menses so that it extends its stay, then it will do so. The punchline hits back once again, that, organisms would rather evolve for resilience rather than production. Furthermore, menopause is seen in communal species (including the few marine species). The organisms that last long enough serve as useful archives for the groups or societies. They assist the young adults and even the children in surviving situations that they themselves survived. In this sense, they perpetuate the existence of the large organism – that is, the population of individuals dependent on them.

Menstruation and menopause thus derive explanations from not only self-organization but from size effects. The strains offered by production increase with age, to the point that it continuously fades as a viable strategy to welcome menopause. To echo Donella 'Dana' Meadows yet again, the greatest input to a system at any one point is the one that impacts the most limiting aspect of the system. For a middle-aged mother menopause is the reasonable option for surmounting the limiting dangers of pregnancy.

By now it should be apparent how size influences decisions, how organisms have become extremely sensitive to size effects and how size pushes organisms to the limit. The faster the organism grows, the faster it approaches its limit. By self-organizing to criticality, it could either collapse (death of the mother, the baby, or both) or circumvent it (give birth of the baby or for the elderly, switch to menopause). It must therefore be acutely sensitive to size effects. Size would then give the organism just enough time to produce. Insensitivity to size effects results in long-standing medical conditions and eminently, death. It might appear that size settles the age-old riddle of the chicken and the egg but I still do not think it does. It could be that the first chicken hatched from an egg or that the mother of this chicken was some creature who had some defect and started laying eggs. As far as I see it, the riddle stands, just like the works of Dana Meadows. For now, let's get into it in the final chapter.

CHAPTER EIGHT

Leverage Points

It's bigger than hip-hop, hip-hop, hip-hop,
*hip – **Hip Hop, Dead Prez***

Throughout the book I have introduced several aspects of systems thinking or systems dynamics. I have one person to thank for that – Dana Meadows.[1] I might have mentioned her one too many times, but it is warranted. She showed me that a system is much bigger than how we understand it, just like how hip-hop is bigger than we perceive it. Hip-hop can tell you a regular day in the life of a young adolescent girl or it can try address some issues that are hardly being addressed by the local or state authorities. Hip-hop can console and at the same time inspire. It is not just an output from a group of individuals hoping to thrive off it. Hip-hop *is* bigger than hip-hop. Dana illustrates that systems tend to have such behaviours. Her summary of how systems work took all my mental models and aligned them into a lattice that I can rely on at any given time, just like a rapper or a singer can rely on Kanye, Pharrell, or even Timberland's *sick* beats. Coming from the medical field, one that for a long time has relied on reductionist principles of scientific analysis and hence a bottom-up perspective, Dana introduced me to a systems perspective, a top-down perspective. I soon learnt to shift my reliance from Occam's Razor – among competing hypotheses, the simplest is the correct one – to Hickam's Dictum – a patient can have as many diagnoses as he or she well damn pleases.

I was, therefore, able to appreciate the 'inescapable' power of a gene-centred view, as Dawkins has explained it, and formulated my theory from a systems perspective. I hoped to paint the picture of both top-down and bottom-up perspectives, particularly using the concept of hierarchies. For instance, a gene-centred view will describe how genes influence behaviour of an organism, but the NK model will show how the influence of a gene on other genes is limited to a small, neighbourly number, thus creating levels of hierarchies that are autonomous and yet dependent on each other in a self-regulative manner. One group will regulate the other, often without their knowledge. The end result will be that each hierarchy level will then generate the system behaviour without any evidence of central control.

In this final chapter, I will honour the insight that I got from Dana by discussing leverage points, which is a treasure trove in systems analysis. The principle behind leverage points has to do with levers – application of effort on one end results in the movement of the load on the other end. In systems analysis, the aim is to apply as little effort as possible to realise a disproportionately large effect. Jay Forester, who is famous for making public these leverage points, insists that they are very often counterintuitive. You will, therefore, find that people put a lot of effort in areas that have little leverage. Systems thinking, through leverage points, shows how such efforts often end up being suboptimal. Henceforth, when a counterintuitive option is offered, it seems so irrational, so … counterintuitive. The word 'growth', for instance, is stressed by many motivational speakers, companies and various economies but growth can be counterintuitive, as exponential increase in Natural Selection shows. Unrestrained growth approaches limits that very often lead to collapse. Controlling growth, therefore, offers invaluable advice for a disciplined form of existence for the person, company, city, or even an empire. Yet, this appears counter-intuitive. All in all, the places to intervene in a system that guarantees it the most sustainable returns espouses the idea of leverage points.

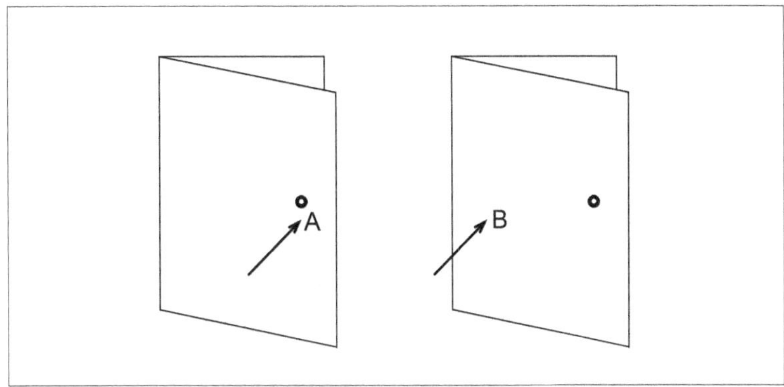

Illustration of the influence of leverage. Pushing the door at point **A** requires less effort than pushing it at point **B**.

From her vast experience, Dana developed a concise list of points. I call them Dana's rank of leverage points. They helped me understand how my theory applied concepts of systems thinking. They also helped me lay down my thoughts lucidly (I hope) about OS. There are twelve leverage points in number and I will touch on how each of them can be applied in organisms, which are systems manifest. In decreasing order of effectiveness, we have:

1. The power to transcend paradigms
2. The paradigm out of which the system arises
3. The goal of a system
4. The power to self-organize system structure
5. The rules of the system
6. The structure of information flows
7. The gain around driving reinforcing feedback loops
8. The strength of balancing feedback loops, relative to the impacts they are trying to correct against
9. The length of delays, relative to the rate of systems change
10. The structure of the material stocks, flows and their nodes of intersection

11. The size of buffers and other stabilizing stocks relative to their flows
12. Numbers – constants and parameters

Before I dive into the twelfth rank, it would be best to demonstrate what a simple system is and how the framework can be applied to organisms. Yet again, I shall borrow the example given by Dana of a bathtub.

When you open one of the faucets in a bathtub, with the drain plugged, water steadily accumulates in the tub. When the drain is unplugged some of the water leaves the tub. Now, if you intend to take a bath, you will have to set a goal, which is, the level of water you find suitable for you to comfortably dip yourself into. You'd then plug the drain and turn the faucet to let water into the tab. The more the water gets in, the closer the system approaches its goal. You can appreciate this trend by observation. As the system approaches the goal you begin to regulate the faucet until it is finally closed and no further water gets in. Your goal has been achieved.

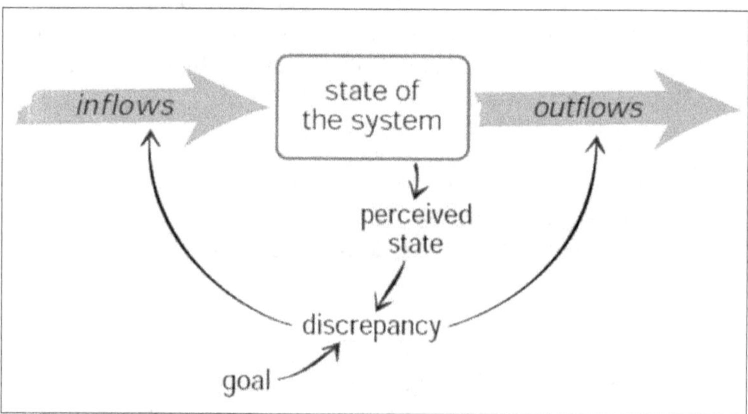

An image illustration of how systems work – it contains inflows and outflows that contribute to the stock (state of the system). These variables are altered and sustained by feedback processes that are shaped by the goal of the system.

Once you have gotten in, some water might spill over. This information forces you to open the drain, until your newly adjusted goal is attained.

All the while you monitor, again, using observation. Once the goal is achieved, you seal the drain. Simple. Let us say you find that the water is running cold. You then open the hot water faucet. However, it takes time before the water heats up and even more for the water temperature to reach your newly desired goal. You would have to regulate the amount of water getting in from the hot-water faucet, while releasing some of it through the drain to avoid spill overs. At this point, two things have happened; some delays have been set in motion and, the system has become complex without your awareness.

While initially you could only achieve your goal by observing the water level in the bathtub, now you have to observe the water level *and* feel its temperature, after some set delay. An overshoot might require regulation of the water level and the temperature using both the cold- and hot-water faucets and the drain. You end up wasting a lot of water due to lack of information but it bears little significance to your goal. In short, systems often turn complex without any of its components being aware of it. Andy Grove likes to call such crucial albeit unnoticed changes as inflection points.[2] Systems are notorious for these unseen moments and are thus irreducibly complex, largely accounting for their unpredictability. Ergo, a system, in itself, is the most unpredictable entity, and thus the most diverse just by mere existence. This underpins the assertion of OS that the fundamental unit of diversity is an organism, appreciable through mere existence.

Let us break down the bathtub into essential components of a system. The faucets are the inflows, and the drain is the outflow. The water that accumulates inside the tub is the stock. As long as the rate of inflow is greater than the outflow, the stock will increase exponentially. Such an increase is facilitated by a reinforcing loop generated by the difference between the inflow and the outflow. You, the monitoring entity, are also part of the system. Your role is to assess the discrepancy between the system and its goal. Therefore, you constitute the balancing loop. A delicate balance between the reinforcing loop and balancing loop determines the behaviour and trajectory of the entire system either toward or away from its goal. As the monitoring entity you are the one aware of the system's goal.

The stock constitutes the state of the system and its alignment with the goal. Its assessment is relatively simple when cold water is used. However, hot water only introduces complexity and delays. The delay further contributes to the complexity. You would have to assess the level visually, the temperature through skin sensation and the delays cognitively. While assessing for one you miss the details about the other. Evaluating the details at the same time becomes a tedious and impossible task. The visual pathway is different from that of skin sensation. I have not even factored in the cognitive assessment. What all these translate into is a complex goal. If a system has a complex goal it will have a complex outcome, often an unexpected one. If it has a simple goal, it will have a simple outcome.

There are also visible and invisible components of the system. The tub, water, faucets and you are tangible. In fact, I am sure some never thought of themselves as part of the system. Systems concerns the relationships between components rather than the substance of the components. More importantly, since systems focus on relationships, its most important aspects are invisible or intangible. The delay, for example, is not evident until it is too late, causing an overshoot in reaction. The reinforcing loops are not evident much the same way as the balancing loops. The goal is also not visible even though it is the driver of the system. Visible components rank low in Dana's rank of leverage points. Invisible ones rank higher. Now that you have a basic understanding of system components, let us now see how they can be related to our topic of interest in this book – the theory of Organismal Selection.

Numbers, Constants and Parameters

The saying goes that numbers don't lie. However, numbers are the least points of interventions. Systems, even those that we think are simple, likely have some hidden complexity about them. Defining a system by numbers has the unexpected tendency of blinding one is to the system's behaviour. The adult human body comprises around 37 trillion cells. There is little that anyone can conclude from this statement. What if we add that the human body comprises 90% bacteria and 10% human cells? That still says very little about the system's behaviour. Emphasis on quantity side-lines the emphasis on the quality. In fact, from the scientific method, it is substituted that an emphasis on quantity captures the quality. Systems, however, evolve past the constants that are attributed to them. Here's are several examples.

When salt is added to pure water, the boiling point increases by some unknown number. The freezing point also reduces by an unknown factor. When calcium passes the nerve-muscle junction it is not known how much is needed to excite the muscle every single time this process is replicated. While some thresholds are established, even those figures are bound by some level of precision that is ignored in the interest of behaviour. The sea creatures are the largest living organisms that we have ever known, and they possess sizes that on land would have posed immense constraints to movement. The buoyancy offered by the large water bodies allows them to evolve past such numerical constraints. When a particle merges with another, numbers do little to describe the behaviour of the system. Since systems concern themselves with the relationships of its components, there is reduced emphasis on measurements and increased emphasis on behaviour. These relationships can be mapped. While the numbers change, relationships can remain intact. In an ecosystem, the fox will hunt the rabbit. This relationship is invariant to the number of foxes in that ecosystem; it has been mapped. The map we have by now, hopefully imprinted, of the organism and its universe, remains robust in spite of every unique particle (organism). Since systems are principally unique, the numbers still matter, but not as much.

One peculiar aspect about numbers stems from hierarchies in relation to the inverse law of large numbers. Firstly, hierarchies grant an organism the chance not to keep complete track of other organisms in various hierarchic levels. An individual organism such as you or me, therefore, has little information about what it is. It has even less knowledge about itself the more the hierarchical relationships it forges. Hierarchies, therefore, reduce the predictability of an organism. From the laws of small numbers, an organism contains unscaled diversity. These two concepts – hierarchy and law of small numbers – illustrate the variability and unpredictability that is necessary for evolution. One of the best ways of being unpredictable is by not knowing oneself in totality. Hierarchies achieve just that. As systems that evolve past set constants and parameters, numbers can hardly be reliable leverage points.

Size of Buffers and other Stabilizing Stocks Relative to their Flows

The size of the buffer can be equated to the size of the organism. Therefore, the stock can be a marker for the stability of a system. Between rivers and lakes, there are more catastrophic river floods than lake floods. The stock seen in rivers is very small compared to a lake. A stable stock can therefore make the prediction relatively easier than a mercurial one. It is, thus, easy to see how this becomes one of the easily testable variables. It allows us to answer the scientific question – is it measurable? If so, can we predict? More than that, it grants us the chance to give qualitative predictions.

Size translates to stocks. Stocks in themselves are delays. There will be a delay when the water fills the bathtub or before the goal is attained. Similarly, in large-sized organisms such as humans, there will be various genetic mutations that do not manifest immediately. Also, the increase or decrease in size is not usually evident when it happens at small scales to large organisms. We can, therefore, recognize that Barrack Obama is still Barrack Obama since he does not change drastically after a haircut. But if we consider the particulate level, the number of mergers and the fall-outs that have happened in the short span of extending a hand to greet him, we will be dealing with different organisms before and after the handshake. The large stock that comprises all the cells of Obama appear stable relative to the flows of small particles that flow in and out of him. Stocks in themselves are delays. We have not evolved a means of capturing these fast changes. If anything, it would only result in chaos and confusion in the setting of long-term goals and as a result, long-term prediction would boarder impossibility. Sensitivity to these size-effects, to the global organism rather than the small particulate organism, grants us qualitative predictions which are immensely valuable in evolution.

A reserved identity, a form of self-awareness that denies annihilation in spite of the particle changes (a form of consciousness), allows for interactions between myself and my neighbour. It also allows for future planning, focusing and learning regardless of the smaller particulate changes. Additionally, an increase in size does reduces the pace of life, a form of delay, when we consider the theory of relativity or efficiency of energy utilization. West clarifies how the terminal units, the points of exchange, are uniformly distributed throughout larger organisms such that aging comes gracefully for the larger organisms. Size is actually a delay and delays create room for long-term planning.

To show the relevance of system behaviour over numbers, consider the size of sea creatures. The inflows being the food that whales take and the outflows being the rate of maintenance. There is little maintenance required in the sea-faring whales compared to the terrestrial elephants when it comes to maintaining hydration status. Furthermore, the size of the whale, relative to the elephant, shows how stocks can increase in size if unimpeded. Sea creatures increase in size past the surface area physically capable of supporting equally sized organisms on land. Constants in the form of numbers are, therefore, the least leverage points. But much more relevant than size in getting the behaviour of the system is the structure that makes up the organism.

The Structure of the Material Stocks, Flows and their Nodes of Intersection

The type of materials that forms the stocks contribute significantly to the type of delay that a system manifests. The relevance of scientific analysis is now evident by asking the question: what is the substance made of?

As a city grows, there is little change in the infrastructure. A single gas station will continue supplying much more people as the population increases. If it gets too congested there will be increasing demands for renovations or newer ones. New roads and infrastructure will have to be made. More material will be needed. However, material stocks are extremely expensive. From an out-of-pocket expenditure getting a donor organ can be very costly. Genetically, recognizing a beneficial genetic switch takes years, or with reference to scale, generations. Even then, the switch might not be all that beneficial, as in the case of Sickle cell mutations, where those with the 'full mutation' (disease) have a tougher time compared to those with a 'half the mutation' (trait). [3,4]

The material that an organism has also forms the means with which it can wade in its universe. The interfaces that interact with stimuli prove to be the most evolving regions in an organism. Take the example of the sensory system. The sensory receptors form the interface between an organism and its universe. They rapidly evolve but at scales that are not evident to the naked eye. Points of interaction double up as the nodes of intersection between and among systems. Within the organism, there are numerous nodes of intersection. For instance, the terminal units such as terminal bronchioles and capillaries. Concerning the material of the stock and the nodes of intersection, there is little that can be done unless there is a complete renovation of the entire system. Yet, even then, the outcome is not a sure bet in any universe. It is expensive in construction, maintenance and the unforeseen downsides they bear.

Little can be done to the material, structure and flows other than making sure the plan is good before the execution is done. We have reasons to believe that genes have instructions which have proven robust for generations now. Dominant genes will always be dominant when paired with recessive genes in what is known as heterozygous states. The much that scientists can do is knock off genes or move a gene from a specific location to another to test its function. It is the organism that then determines how alteration of these instructions fits with the pre-existing ones. Herein lies the relevance of the material and the nodes of intersection, which contributes to the arrival of the fittest. In this process, the organism generates its own knowledge as it traverses its universe. Even more robust than nucleic acid molecules are the elementary particles, for which we have little power over their behaviour. Three quarks end up being more stable than two quarks. That three quarks would have been stable could barely be predictable from the highly unstable two-quark relationship. As Brian Goodwin emphasizes, composition is not sufficient to predict form.

Additionally, size is largely a function of the material that makes up the system. Increasing the size past the constraints will result in collapse or production. The human will have the reproductive system. The sandpile will have the sand and its relation to its constraint. The constraints that size bring are largely a function of the materials the organism is made of. As such, the offspring produced will, therefore, be to a significant degree a product of the material the parent had. It should contribute to the similarity in appearance and behaviour. As for the poor, increased production acts as a wise strategy in bringing more food. For the rich, more children could mean more constraints. Size, therefore, is not just important but the structure of the material that makes up this organism is. Tying together the size and the material generates the types of delays seen in systems.

The Length of Delays Relative to the Rate of Systems Change

Delays are inevitable in systems. From the coding of information to its execution, there is a delay. Our sensory perceptions are delayed. Over time, however, they have been fine-tuned, to keep up with an acute sense of awareness of one's surrounding – one's universe. Acute responses can save lives much the same way as they can destabilise a system. A fast birth process can save the life of the mother and the child. Arrhythmias, on the other hand, can be fatal.[5] Sensitivity to size effects makes up for these acute responses. Destruction of this sensitivity results in delays which can be catastrophic. Prolonged labour can end up in the death of both systems, that is, the mother and the child. This is an example of a delay relative to the system's change. There is systemic evidence of the baby increasing in size but a delay in efficient labour response. Surgical intervention occasionally rescues patients in such situations.

Awareness of systems delays allows one to properly plan and act fast before there is an overshoot (as in the case of preeclampsia, that is, hypertension due to pregnancy) or a collapse (maternal death) of any system.[6] Since organisms intend to avoid annihilation, that is, collapse, they would ensure that they effectively execute mechanisms for minimising such delays. The uterine muscles would then have powerful contractions to expel the baby. Molecular signals have created axes that execute these strategies with shocking precision. The right materials are essential to officiate these strategies but if coupled with delays end up in futility. Thus, delays become more important than just the materials of the systems.

Delays are also inevitable in the growth of organisms, and there is little that we can do about them. An elephant will grow at its own rate, as it pleases. The population growth will grow at a certain rate which we can never capture accurately. What we can do is increase our sensitivity to delays, in the way our senses have, and fine-tune them to disasters before they occur – disasters such as complicated multiple gestation (twins), which might call for surgical intervention; disasters such as increased population explosion in cities, which will lead to the mushrooming of slums, increased unemployment and crime rates; disasters such as increased global temperatures which might lead to the death of many ecosystems. Sensitivity to these delays demands an astute form of balance. A balancing loop, which keeps things in check, thus, bears more relevance than the delays.

The Strength of Balancing Feedback Loops Relative to the Impacts they are Trying to Correct Against

Balancing feedback loops are often neglected yet their roles are vital in systems' resilience. Their central role is regulation. They will regulate the temperature of the Gaia, in spite of the extremes at the poles and the tropics. The Daisy World model captures this beautifully. In this sense, the autonomy of an organism trying to increase in size, will be regulated by another organism also trying to increase in size. In hypercancers, the secondary cancer ends up regulating the primary one. It also explains why doctors would still insist on the 'normal' body temperature range in spite of the ability of the hands and feet to withstand extremely cold temperatures (the body's poles). In fact, the temperature taken is often at core centres, such as the armpit, the rectum or the mouth (the body's tropics). The prevalent theory of thermoregulatory properties of the hypothalamus is also dependent on balancing feedback. The balancing feedbacks are therefore not able to be detected beforehand in an organism. They constitute one of the most relevant albeit invisible components of systems.

Balancing loops are generated by the very organisms that seek to satisfy their own goals. Since their goals are always capped, they will continue to act towards it, with one organism evolving to regulate the other. There will be co-evolution. It is therefore important that co-evolution happens or else one organism will increase in size to the detriment of the entire system. For instance, an uncontrolled increase in *Clostridium dificille* in the gut of the human being will result in disease, that is, pseudomembranous colitis.[7] If the goal of an organism is not capped, it will likely lead to its own demise, as predicted by Natural Selection.

These feedback loops also underscore our assertion that organisms evolve for resilience rather than production. They are, therefore, not obviously seen or their role not easily or immediately appreciated when situations are normal. However, their presence proves vital in the future and often in unusually hostile situations. For instance, when one is hiking up a mountain the body's core starts shivering and the skin develops goose pimples. These are independent mechanisms that end up having synergistic effects. Furthermore, they enable an organism that has always lived in low altitude regions survive during the 'unusual' high altitude situation such as the climbing of mountains. In normal conditions, they might appear to be costly. But during 'abnormal' conditions, they become pivotal to survival. In ensuring organismal resilience, balancing feedbacks and self-organization are invaluably linked.

From the perspective of saving an organism from unforeseen dangers, balancing feedbacks highlight the similarities between the 'living' and 'non-living' organisms. So, does a rock have the same features as a human being? Yes. Firstly, that it has chemically organized components shows that is it is not just random – the aggregation of those particular atoms in *that* particular location does not necessarily strike as purely random. Secondly, the components of the rock are not aware that they form this large entity called a rock. Insulin is not aware that it forms this big entity that is called Thomas or Samantha. Often enough, the entity also does not know its constituents completely – the rock does not know that it has formed several covalent bonds; nor does Thomas know that insulin is produced in equal quantities as another molecule called C-peptide.[8] We do not know around 90% of the cells that help us survive (bacteria and viruses), yet their existence is necessary to our survival. This all happens because of the hierarchical systems seen in organisms.

These balancing feedbacks are latent, and are hardly ever used, unless an emergency situation arises. The mother needs a latent organism namely, the uterus, which is sensitive enough to facilitate the pushing process once the limits for the growth of another organism, the developed baby, nears. Wanyama has an acquired immune system whose secondary role remains latent most of the time. The defences kick in when another organism seeks residence in the building, forcing Wanyama to produce gastric fluids, antibodies and other killer cells. A sand-pile has a critical point beyond which it cannot rise any higher. It highlights the relationship between an organism and its universe. Since the universe is hell-bent on destroying the organism, it acts as the regulator of the existence of an organism in whichever form. Even the role of the universe as a regulator in the size of the organism is latent but becomes evident to the keen observer.

Balancing feedbacks ensure that organisms evolve for resilience. The rock did not anticipate that sometime in the future a mason will try to break it down, but the latent balancing feedback loop was summoned when that hammer forcefully landed on it. Even though the mason often breaks rocks, it takes a tremendous amount of energy to break them. This particular organism is no different from the biological organism that also tries to avoid annihilation. Now, let us say the rock was so hard that the mason only managed to chip off a small portion of it. If Wanyama managed to walk by and witnessed the mason's futile attempts on day one, and passed by on day two, he is still bound to recognize the same rock, even though it was chipped a little bit. While the rock is not the same as it was yesterday after the mason's unyielding attempts, it appears the same to Wanyama. While Olunga has shed off some of his epithelia in various parts of its body and lost some electrolytes and water, he is still recognised by his teammates when he goes back to the football field. A shaved beard does not change how he is identified by the team. Here, size has acted as the balancing feedback that reassures the image recognised by an individual or other individuals in the event of small changes such as a morning shave or an afternoon facial. The evolution for resilience allows an organism to maintain its general outlook despite the small-scale changes not being observable to the naked eye. To get here, however, the organism has to grow in size. This is achievable through the next leverage point.

The Gain around Driving Reinforcing Feedback Loops

Even more important than the balancing loops are the reinforcing feedback loops. They are sources of growth, explosion, erosion and collapse. As a point of intervention, it makes sense to reduce the growth rate which will lead to eventual collapse rather than hope that there will be another organism that will cap the exponential increase. It makes sense to slow down the car rather than complain to the company to create more efficient brakes. It makes sense to reduce the rate of hydrofluorocarbon production rather than push for an efficient way of reducing them from the atmosphere – a rather difficult task. It then explains why the elephant can live for as long as it does since its growth rate is much slower than that of the blue whale. Even trees that last for centuries use this concept.

As the engine that drives Natural Selection, reinforcing feedback loops usually lead to collapse through exponential growth. Continuous production of organisms leads to predictable collapse once the constraint is reached but cannot be circumvented. An example of a constraint is the carrying capacity of a particular habitat. Those that manage to mutate or adapt or survive would form the new organism or, according to Natural Selection, the birth of a new type of species. Organismal Selection asserts that an increase in size enhances the survival capability of the organism, which can comprise individual members of a particular species. In fact, the most efficient means of size increase, eusociality, arguably guarantees survival more than the other organisms, social or otherwise. E.O. Wilson considers it the final destination that organisms gravitate toward. I would add that cities are also one of the most efficient organisms that societal organisms gravitate towards even though they consume resources alarmingly. Stars would then wish to mutate into black holes as black hole efficiency is unparalleled.

Here is a strange structural organization to ponder over. In eusocial organisms (*eu-organisms*), picture the queen in the middle and the rest surrounding and guarding her. Cities are often surrounded by smaller towns, mimicking the eu-organism. Large celestial bodies surround black holes, just like the eu-organisms. Coincidence? Maybe, maybe not. Size, while important, has to have efficient structuring to survive in its universe. Developing efficient ways of doing so is, therefore, an evolutionary task that all organisms go through. Such a structure mimics the fractal branching, which enhances the growth of the ant colony, city or even galaxies. Such explosions explain the production of new ant colonies, expansion of cities and mushrooming of shanties and slums and the expansion of galaxies. Their growth nevertheless approaches efficient balancing if controlled. If uncontrolled, the universe acts, swiftly. If you are an organism, it serves to reduce your rate of growth rather than have the universe take action.

A systems perspective showcases how reinforcing loops clarify the assertion that the bacteria or virus does not attack you, you set up the conditions for it to flourish within you. And radical oxygen species do not attack you – you set up the conditions for it to flourish in its self-reinforcing loops. In such situations intervening by reducing the gains of the reinforcing loops is a much better strategy than hoping that the balancing loops will kick in. Reducing population growth becomes a better strategy than hoping that some disease will emerge in good time and reduce the numbers. This and many more achievements demand an effective means of information flowing throughout the system.

The Structure of Information Flows

Information flows are the invisible forms of communication between and among organisms. For the system to remain intact, there needs to be a flow of information. It happens when there are channels for energy that permeate the organism. In metal, the electrons can move throughout it to transmit energy. The atoms can also do the same within the same metals by vibrating. Bacteria, which I reckon are a single organism, have a means of detecting each other by 'talking' to each other through molecular signals, a phenomenon known as quorum sensing. In mammals, the circulatory system, neuroendocrine system, digestive system, and any other system, literally have material flowing through them, constituting a means of information flow. In molecular studies these systems, even at the cellular level, are said to have cross talks.[9, 10] Essentially, for a system to be intact, there must be a form of communication, which is possible through energy transmission. They have to be thermodynamically open.

With reference to annihilation, information is present when there is an existence of a particle. As long as a particle exists, we can establish its position, mass, charge, momentum or spin. In the absence of existence, there is no information. The further one gets from annihilation, the more information one possesses. As such, much more information is derived when a system is far from maximum entropy than when it is closer to it. But every organism is present in its own universe. So, the information will always be different. It then makes sense to say that we do not talk about what we see but only see that which we can talk about. When measuring the same object, my measurement will differ from yours, even if we are using the same measuring tool. It is the reason an average of several measurements of the same object is often recommended in such situations. Another way of putting it is that it is easier to generate than to trust information. It validates the map and, with regard to information, the organism as the fundamental unit of diversity.

Lovelock noted that the Gaia was in a state of extreme disequilibrium, far from maximum entropy. It was therefore teeming with information. The other planets are close to equilibrium, and are, therefore, closer to annihilation. They bear little information. This does not mean that there is *absent* information. Existence in itself is a form of order, and hence a particle bears some form of information. It brings us back to the idea of unscaled diversity. At its core, *unscaled diversity* means that an organism can attain levels of order that are unpredictable. From an initial world consisting of only elementary particles, it is close to impossible to predict the future existence of the snowflake or even the dinosaur. These are structures that could not be predicted before their emergence and subsequent familiarity. Even the existence of a particle in its unique spatio-temporal niche could not and cannot be predicted, and it still cannot be predicted with absolute accuracy (whatever that is) to date. Existence and its distance from annihilation determines the degree of information that can be derived from the organism. The further it is from the cessation of existence (however that can be measured) or thermodynamic entropy (which can be measured), the more order it possesses and the more information it bears.

The order does not necessarily have to be because of universal pressures to annihilate it. It would not be *unscaled* diversity if the level of diversity matched the level of inhibition of the universe. That would make it scalable. The diversity that an organism wields has to exceed the inhibitory force of the universe for it to last as long as it does. It has to be bold and unique first before its universe annihilates it. The organism, therefore, is shaped and shapes itself in dimensions that surpass the idea of adaptation and hail itself as a new form of order. It does this by evolving for resilience, through hierarchies. Stable hierarchies allow for an efficient flow of information without causing 'information overload'. The heart only needs to know the pressure at the vascular system, but not the entire structure of the red blood cell. For a long time, we have never known the role that the 'animate' sea planktons play in regulating the temperature of the earth, let alone the 'inanimate' polar ice caps. The right flow of information determines the balanced maintenance of feedback loops and minimises delays in spite of the stocks that organisms comprise.

In a harsh universe, little room is given for experimentation but that is the only room available to make the jump for a bold strategy never used before. Risk-seeking behaviour appears rewarding even though the chances are cosmologically slim, relative to the exceedingly high probability of annihilation. This would mean that occasionally, organisms can attain success. With bold strategies, such as the formation of viable mergers, they postpone their annihilation. Since every particle or order of particles represents information, the more the mergers, the more complex an organism gets. With symbiogenesis, particulate or biological, through the process of self-organization, there is the emergence of a new organism, statistically speaking, with new properties and a consequent increase in size. The more complex they get, the harder the universe finds it difficult to annihilate, thereby increasing the gap between the present state and the maximum state of disorder of the universe. So, the universe will continue to gravitate towards annihilation but as long as there are organisms it will not happen. While the world strives to be boring, the existence of any particle counters this drive. The universe and the organism are tightly coupled in this journey of evolution.

There is an important discrepancy that I have to delineate. The idea of the gene is understood differently by two fields – molecular biology and evolution. Molecular biology considers the gene to be a component of the DNA but the evolutionist considers it as any piece of information that can be passed down through generations for long enough for it to be considered for Natural Selection. When I mentioned that twins are different in spite of their similar genotypes, the gene-centred view, according to the evolutionist, is fairly acceptable. It is the gene as understood by the molecular biologist that weakens since it does not explain why the two are different besides the idea of *developmental noise*. Developmental noise comprises the errors of genetic replication that are also evident in identical twins. It is, therefore, easier to say that the *DNA*-centred view weakens in insisting that the identical twins are 100% similar genetically. However, the *gene*-centred view remains fairly stable. The DNA and its components do not form the gene, the gene is a *code that is archived in the DNA* and in other organisms, in the RNA. It can be easily associated with the DNA since the information is passed down through generations after 'reproduction' but this should not be the reason for the misunderstanding. The gene and the DNA are two different things. Since genes are considered pieces of information; they hold systems intact. However, from the works of Stuart Kauffman, a single gene will only influence a handful, to make a stable and efficient hierarchy that is the genome (DNA). A stable one would allow for a measured flow of information and a progressive balance among gene networks as an organism increases in size.

However, there is a subtlety that weakens the gene-centred view much the same way as it does the DNA-centred view. Since every particle and thus every organism is unique, the information they bear is also unique. Genes might be taken to be the same so far as they are stem from particles, but they are different. They may be *similar* but they are not the *same*. They are different for the simple reason that particles are different. The example of the haemoglobin molecule is often cited to explain the fidelity of the replication process and maintenance of the gene instruction for generations. However, genes have several redundant repeats that do not manifest phenotypically. A small mistake in the process of haemoglobin formation will hardly be noticeable. Thus, surface evaluation can give the impression that the replication is perfect. The fidelity of 'replication' is high but it can never be perfect. Nor does blood transfer from one identical twin to another happen seamlessly. There will always be some resistance although mild in severity that follows from the difference in molecular structure, which stems from a difference in the genes.

The very idea of errors of replication is taken to be the engine of evolution. Such a process can appreciate *errors* only when we accept the idea of high-fidelity reproduction. However, when we shift from reproduction to production, they are no longer errors. They are bold steps, each bold enough to assert its existence. At the same time, the process maintains the engine of evolution since each production results in a unique organism. Furthermore, shifting from reproduction to production immediately changes the perception of a single gene, lasting in a gene pool after 'reproduction'. It only lasts as long as it exists. Reproduction does not happen, only production. This is the small leak that has the potential to sink the great ship of a gene-centred view. Thus, every organism will go about its universe collecting different types of information and some might be archived in what we have come to appreciate as genes encoded on DNA or RNA.

At this point, I would want to poke your brains a little bit and go back to some speculation that I had earlier discussed and which warrants investigation. If genes are archived in the DNA or RNA, why should it just start with the DNA or, more specifically, with the RNA? Was there no information before the RNA formed? A step back from the RNA world to the pre-RNA world shows the logical potential for information to also be archived in the molecules that preceded the RNA. The RNA and the DNA should therefore have formed through some form of order, such as spontaneous order. It might have formed through self-organization, as Kauffman posits. Particles seek conjugate particles, merge and have emergent properties that ensure their stability, as in the case of the RNA or the DNA. Now, we had mentioned that self-organization has the unique property known as power-laws, which allows particles to withstand perturbations. That said, the particles preceding the RNA, which existed in various forms for billions of years had to be more stable than the RNA or the DNA, and they were. The RNA proved its mettle by also manifesting its stability. However, it was hardly as stable as the DNA, so the DNA dominated, again, following the process of self-organization. Mergers guided this process. Viruses and bacteria then emerged and became easily stable and continued to be for a long time before the emergence of the archaea or the eukaryotes. Mergers guided these steps too. These are stable hierarchical steps that mark the history in the evolution of information. The ease with which RNA forms proteins in the presence of amino acids could just be self-organization manifest, where particles seek mergers and end up as dynamically stable forms. Self-organization, therefore, tells the tale of information flow and how it has been archived all along. As much as genes have been at the centre stage since Hugo de Vries, I tend to think that some form of information could be archived in every particle, and not just the DNA. The DNA only has the idea of replication through the 'reproduction' process. But after doing away with the idea of 'reproduction' and shifting to *production*, I would argue that particles also have archived information. This would make every single particle an organism and, thus, do away with the question of the significant leap from 'dead' matter to 'living' matter. In line with Kauffman's idea of phase transition, this is my alternative theory of how organisms emerged and operate. It nevertheless warrants much more thought analysis and experimentation. This *modus operandi* should then take us to the next leverage point.

The Rules of the System

The rules of any system define its capabilities. According to Mendelian laws of heredity, genes are segregated, passed down to offspring, and merge once more, one gene from the mother and the other from the father.[11] However, other forms of non-Mendelian inheritance have been documented. Rules exposed the Mendelian laws to scrutiny and tests, which showed that there are other organisms and diseases that follow other rules.[12, 13, 14, 15] In the same spirit, we have already discussed how there is no such thing as reproduction. The closest we can ever arrive at such a term is in asexual reproduction, where they are *genetically* (almost) identical. But even then, the resultant organism is different, since they contain different organisms who are products of an imperfect attempt at reproduction. Furthermore, produced organisms occupy different spatio-temporal niches hence bearing different information.

Organismal Selection introduces other set of rules particularly on production. The rules satellite around to size constraints. For instance, the qualitative rule that governs systems is that they would tend to avoid annihilation. Another closely related quantitative rule is that organisms, biological or otherwise, would tend towards an increase in size but this will be limited by the size constraints. Organisms will self-organize to self-criticality. If the organism is unable to eliminate this constraint, it is bound to collapse. If it lives within this limit, it will progress towards certain annihilation. If it eliminates this constraint, as in the case of organismal 'reproduction', there will be production of new organisms. It calls for the ability of a system to self-organize, which is the next leverage point. It will become clearer how rules constrain but self-organization does the opposite – it frees.

Another rule is that as long as there are particles, organisms exist. This paragraph might be a tad technical but I shall attempt to explain it in prose form. Should you fail to follow, there is little that is lost from the main argument by OS. From Einstein's theory of relativity, as long as there are particles that contain mass, there is time (and space) for evolution (mass almost equals frequency). In the absence of mass-bearing particles, there is no notion of time or space and, therefore, evolution takes a turn for the bizarre. There are only massless particles such as light, which exist. In a universe that lacks an idea of what size is, which is dictated by mass, all the massless particles present allow for the possibility of the emergence of future forms of organisms. According to OS, they are particles and thus, organisms. Roger Penrose further asserts that this forms the transition from one observable universe to the next. Bearing in mind that every particle is unique, we can argue that the particle diversity allowed for life to bounce back and mass-bearing particles to form. We move from one bounce to the next.

All this is possible through self-organization. For there to be an unpredictable future outcome in evolution, there should be more of free, as opposed to limited, room for experimentation. The space of possibility should exceed the present reality. The mere existence of an organism runs contrary to what the universe prefers, so why abide by stringent rules? Self-organization should therefore rank higher than the rules.

The Power to Self-organize System Structure

Self-organization *capability* is the space of manipulation that an organism has within its bounded self without significantly changing itself. The atom has the electron and nucleus, and the space of manipulation lies first with the electrons around its orbitals, and secondly with the nuclear components. Electrons can swim around the orbitals without significant change to the atom. The nucleus contents can shift in position with little change to the atom. Combined, these capabilities can create room for mergers with other particles. Noble gases such as helium, neon and radon have remarkably reduced *ability* to self-organize and hence might not be considered as 'exciting' as the other elements. The other elements that have heightened self-organization capability 'excite' the particle physicists and chemists. They are, therefore, able to evolve in recordable steps compared to the fairly inert noble gases. The noble gases can be compared to the junk DNA (introns) found in organisms, which bear little excitement compared to the reactive elements that are similar to the meaningful DNA (exons). Organismal Selection asserts that these entities are not just junk or inert, but forms of order that have managed to avoid annihilation. They are spontaneous, coherent, invariant to many universal perturbations and evolve significantly through hierarchies just like any other organism.

Hierarchies that persist acquire stable forms that assert competence but since they minimise the amount of information that organisms have to keep track of, they reduce comprehension. Organisms, through self-organization, are then capable of attaining competence states without future comprehension of these states – a form of competence without comprehension. Kauffman and his team have been able to show how the transition phase, between rigid order (as in the case of ice) and extreme chaos (as in the case of water vapour), lies the excitable media that constitutes living organisms (as in the case of water). This state between extreme order and chaos captures entities at the edge of chaos. What OS shows is how the other states can still be considered as organisms, even though they may not be as exciting or excitable as the systems at the edge of chaos. They still exhibit variant forms of competence without or with minimal comprehension.

Self-organization can happen through the synergistic effects of the different organisms in different hierarchical systems. The ants, once they discover that some caste of workers are absent, can progressively re-organize and re-allocate themselves to fill the gaps using the information derived from the pheromones they produce.[16, 17] Similarly, through positional information, a cell can identify its position and change its form. A city would continue to build its businesses and infrastructure even after a bombing or any other similar attack. The traffic would re-organize itself based on the different available routes to a place due to the various nodes of intersection of information. Trying to avoid the threat of being stuck in a traffic jam would then lead a city to self-organize by seeking alternative solutions. The ability to use what is available to maintain the structure and alter the available materials to avoid annihilation constitutes self-organization. It is the creative element of a system. Organisms, thus, are active and creative in generating strategies to postpone their doom. On the contrary, Natural Selection does not create, it eliminates. Creation is significantly organismal. It is the emergent property that stems from the autonomous organisms, contributing to resilience and dynamic stability. It manifests itself when an organism senses increased deviation from its goal. The goal will then guide the organism to re-organize and minimise the discrepancy between the present state and its desired state. It explains why the goal ranks higher than self-organization.

The Goal of a System

The goal of a system is very interesting in the sense that it drives all the other components of the system below it – self-organization, rules, information, and the feedback loops. In our theory, we have stated that the goal is to avoid annihilation. Phrased in the negative sense, it is different from how we have always known goals to be. Goals are often positively stated. When a player scores a goal, it is a plus for the team, not a subtraction. In Natural Selection, it is to reproduce more efficiently if not more 'efficient' offspring. Nevertheless, framing a goal in the negative gives an organism some room for flexibility, for dynamic exploration. It does not narrow the rules governing the organism. It appreciates that an organism will even break its stated rules provided its present efforts help it avoid annihilation.

A cancerous organism will break the rule of the archetypal ways of generating energy and create its own way. For female human beings will self-organize and shift from a menstrual phase to a menopausal phase. Bacteria or plasmodia do multiply inside a cell such as the red blood cell, until it goes rogue and causes the cell to erupt even though the cell housed it for a while. The female elephant will opt to regulate its growth and that of the organism in its uterus.

In a hospital setting, the goal helps the patient identify what it considers 'not-living'. Using a ventilator might be horrible to one, and in another, appealing. Blood transfusion for a follower of Jehovah's Witness is close to annihilation as they see it, and they would want to avoid it as much as possible. 'Do no harm' is the goal of every doctor, which is also set in the negative, to give room for a wide variety of options for the doctor and the patient. Consciousness, a state that is not just limited to the brain, avoids annihilation by asserting its existence every time somebody calls you by your name or asks if you are self-aware. This response is a concerted effort of the brain, the respiratory system, the circulatory system, the musculoskeletal system and at the molecular level, the immune system. Their goals overarch and in a brief moment, are cogently aligned. You can then respond that you are present at that moment. Pertinent to this is how each has diverse roles that work to keep the individual running. Diversity yet again enhances self-organization capability, only when goals are aligned. Humans, therefore, can still establish a healthy relationship with Gaia by overarching goals, since each does not want to be annihilated.

Stringing the same thought, only when the goals are aligned do two or more original organisms develop into a new emergent organism. Chemistry teaches us that hydrogen bonds are slightly stronger than

Van der Waals forces but weaker than covalent or ionic bonds.[18, 19] But in spite of the use of the word 'bond', the molecules are usually distant. Physics makes it even more elaborate – the distance between the proton and the electron is substantial, such that if a true scaled version were to be formed it would be larger than the diameter of over ten solar systems![20] However, the connection between the proton and the electron is still evident. The space that a tree and another tree have between themselves might create the illusion that they are separate and distant, as separate single entities. They are, however, related, if not through gravity, then through the same wave of carbon dioxide they take. If not through that then through the sunlight they take. If not through that then through the water they sip using their roots. Evidence collected by disparate fields of science shows that by these different forms of relationships all particles are connected. And there is more. Just as organisms are emergent, so are the connections. Each new connection potentially creates a new organism whose presence can be tested by subjecting it to some form of credible threat. Lovelock is justified in calling a forest ecosystem an organ of the Gaia, much the same as a human being consisting a menagerie of bacteria and viruses, known and unknown, that constitute their normal wellbeing. And if calling on to physics might seem far-fetched, then yielding to biology might help clear things up. Each of those trees feed from a common source of carbon dioxide and a common source of water and take in a common source of sunlight. The only thing we visualise is how separate they are standing independently above the ground and not how they care connected by their source and their sinks. They are just like the hair strands on man's head – although they are different, they are connected. By establishing a connection, two or more organisms create an emergent organism. These connections can only happen if some facet of their goals overarch. However, the sure-bet way of confirming the presence of an organism is not just through mergers, but by subjecting the organism to some form of threat.

The goals of systems can be so powerful that it produces system states that seem impossible to revert from. For instance, if my goal is to see my friend suffer, and the same goal applies to my friend, the situation would escalate very fast into a vicious reinforcing loop. For one gang, annihilation could be as simple as being disrespected by another gang and disrespect can come in many forms. Not responding to any such form of disrespect is equivalent to being considered a wimp. Gangs would therefore always attack each other in a myriad of ways. Different ant species live in such a world, of hostile escalated attacks. It applies to any organism that has a goal that is not well defined. In the case of OS, having a goal such as avoiding annihilation explains why such escalations happen. It ends up being what game theory calls an infinite game or an iterative game. Schools would insist on producing more star pupils, businesses seek methods of increasing market share with better products, and economies strive to prove that they are better than others using metrics such as the Gross Domestic Product (GDP). In this way, the goal to avoid annihilation show just how two separate organisms can influence the evolution of each other over generations. However, as stated earlier, such reinforcing loops tend to blow up at some point if not controlled. If one organism cannot match the other's tenacity, it can end up being outcompeted and completely annihilated. The remaining organism would then end up occupying this niche until it reaches its carrying capacity. The goal of avoiding annihilation through increasing in size continues past the point of annihilating a competitor. A keen observer will notice how the map of an organism and its universe is consistent in all these dynamic relationships.

With reference to the GDP, the goal does not even include the quality of life of people – only that the GDP rises. This is an example of an archetypal problem identified in systems dynamics as: seeking the wrong goal. Remember, goal aligns the system with everything below Dana's rank even if it harms the constituents of the system. It is the reason I insist that the goal of the organism be the tendency to avoid annihilation, as organisms will do anything to achieve it – it will self-organize, fuse, seek alternative mergers, move (kinetic theory of matter), increase in size, eliminate size constraints, store information, create a form of self-awareness known as consciousness, steal, cheat, kill, create dubious metrics, wipe out a big chunk of a particular race, invade ecosystems, develop large plantations, continue polluting resources, and in the process justify every single action to avoid annihilation. It is using this point where my theory differs from that of Lovelock and Lynn Margulis – organisms, even the Gaia, have a goal. The Gaia will wipe out the human race just so it can maintain temperatures it has worked for millions of years to sustain in light of the alarm caused by global warming. The Gaia has managed to maintain this goal in spite of the increase in solar output by over 25% since life began on Earth. I do not think that humans will be the ones that stifle its long-earned efforts. Insisting that a particle or even a planet such as the earth has a goal means that I am asking people to consider a different paradigm. Paradigms shape the goals of systems. It explains why it is higher than the goal of any system.

The Paradigm out of which the System Arises

I consider the greatest evolutionary achievement to be that which is considered so commonplace, so unperturbed by different environmental conditions, that it is taken as a normal. Such entities might even be considered dead. For instance, for particles, such an evolutionary milestone is acceptable as the norm so much such that nobody even thinks that they could be sharing properties similar to biological organisms. This is the power of paradigms. According to our biological understanding, the prevailing paradigm dictates that an organism has to have a metabolic process and 'reproductive' capability. To accept a new theory, paradigms have to be created, changed or abandoned.

Take the example of noble gases. They have been able to avoid annihilation so much that they are almost considered irrelevant. Almost. They cannot be thought of as organisms due to their 'inert' properties. However, they have managed to obtain an envious resilience for so many environmental conditions that they ought to be giving us classes. Despite their 'inert' nature, they still have properties like that of any other organism. They, arguably, more than any other family of elements, show that organisms evolve for resilience. The other elements that are always reacting are considered 'active' since they lack this *noble* stability. They are only reactive since they possess the goal to avoid annihilation and a heightened self-organization capability that is fuelled by this goal.

Similarly, the ability to form groups, couples, families, neighbourhoods, counties, cities, states and even armies is taken to be so normal that their relevance is hardly linked to evolution. Were it not for pioneer scientists like E. O. Wilson, the tendency to form groups would have eluded many evolutionists. The same tendencies stretch to the atomic and sub-atomic scale but such evolutionary leaps go unappreciated. Elements 'react', but doesn't this process result in compounds? Nuclei 'react', but don't they form new elements? Wouldn't a better term be that these particles 'act'? Or could it be that they are still 'reacting' to their harsh universe? I cannot decide – I'll leave that to you.

This, again, is the power of paradigms – they are able to prevent thought. Garrett Hardin has constantly reiterated the ability of words to prevent thought. Paradigms facilitate just that; they can allow for the exploration of a whole new world while preventing a thousand others. They create what Charlie Munger calls a man-with-a-hammer tendency. But a powerful hammer is a terrible saw. Belief in the paradigm of Natural Selection and its understanding of reproduction can prevent a scientist from believing in production and the theory that I tried to elaborate in this book. I am not insisting that my theory is true. We cannot know whether a theory is true or not. I am merely trying to persuade the reader to consider the possibility that there can be another explanation for evolution. Furthermore, considering alternatives distances one from our inbuilt bias, allowing us to reap the benefits of diverse perspectives. It is mathematically proven that a diverse perspective is more accurate than a single perspective. With this in mind, accuracy should be the goal of science, not a constant struggle of who is right and who is not. In the interest of achieving accuracy, we should not be bothered that we are wrong or celebrate by chest-thumping that we are right but that together humanity is steadily gravitating towards accuracy. It is from paradigms that such goals are set. I went through these same steps. As I started thinking about this theory, I was plagued with the idea that I might be wrong. Every scientist hoping to develop or discover something big knows of this fear. In the evolution of its existence, that is, the theory of Organismal Selection, I learnt to shed off that fear in the interest of accuracy. This will become clear in the next leverage point. It introduces clarity and clears the unwarranted burden of being right, paving an exploratory path leading to accuracy even if it is marred with failures.

Perhaps a good clear example of how paradigms prevent thinking is seen by comparing these two perspectives. From Natural Selection, if you cannot reproduce you are virtually irrelevant. From this perspective of evolution, *you* have a problem. Organismal Selection dictates the reverse – that from the point of *any* organism, the universe is the problem. It tries to annihilate while the organism tries to continue existing. Having these two perspectives allows one to appreciate nature bi-focally and, hence, in a much richer way.

Paradigms further highlight that the higher the leverage point the higher the resistance. Take, for instance, the development of the normal curve, what is colloquially understood as the bell-curve. When Francis Galton described this pattern of distribution, the bell-curve distribution, it was believed that this should be the distribution of virtually all data.[21] The bell-curve, however, was not able to explain the power law distribution, which also stems from random behaviour. At the basic level, what is often taught is the normal distribution, the bell curve. The other distributions, such as those that Benoit Mandelbrot describes, are dismissed as outliers or pathological entities since they are 'not normal'. These 'pathological' distributions have been resisted for a long time and have only begun to gain credence in the last couple of decades.

I like to entertain the thought that OS describes why these variables follow power law distributions and the normal distributions. Power law distributions display the interdependence property and normal distribution display the independence property of the variables of one's interest. Similarly, an organism would want to exist as it is, independently, but it is inextricably linked to others – it is interdependent. These two distributions are inevitable due to the inescapable relationships between organisms and their unyielding drive to assert their independent existence. The interdependence would then explain the various catastrophes seen in evolutionary history such as the Permian-Triassic extinction events. These are events signalling production after organisms had self-organized to the point of self-criticality.

I would add that the resistance to paradigms validates my single test used to detect the presence of an organism – exposing it to credible threat. By insisting that someone change their ideas, you are threatening them, that they might have believed something that might not be as close to the truth as they might have thought. It would then explain why the particle would want to avoid annihilation since the paradigm it has always wielded is used to shape its goal. An organism believes it exists. It will resist anything that thinks otherwise. Taking this belief, it will shape its goal – to avoid annihilation. Paradigms will then dictate and always rank superior to the goal.

Paradigms further explain what Dawkins has called the tyranny of the discontinuous mind.[22] That is, essentialism. Extending his argument, there is a conventionally accepted rule that we can create a slice between the living and the non-living. We then seek features that support our arbitrary split and, hence, our arbitrary boundaries in our various disciplines. Metabolism, DNA, RNA, the cell, can then be used to distinguish living from non-living entities. Considering man's creation from other forms of creation is used to split between the natural and the artificial. There is a human need to split things into discrete entities in spite of the more predominant continuity of nature. The species *Homo sapiens* did not appear out of the blue when a mother gave birth. The very first was an organism. The need to develop groups such a species is an arbitrary boundary that allows for the studying of groups. Sadly, this thinking has the potential of blinding one to the continuous nature of existence. At which point do particles, considered dead, turn living? It becomes ironic how biology has the power to consider a conglomerate of particles living and another set dead, even though they all began as 'dead' entities.

Strictly speaking, discrete boundaries hardly exist in nature. This is the case despite various intrinsic group abilities and behaviours that can result in unilateral intra-breeding tendencies and, thus, the emergence of closed systems leading us to create a biological entity known as biological species. Thermodynamics accepted that nature has open rather than closed systems. Extending the tenets of the same argument, I have used evidence from various disciplines since academic boundaries are also largely arbitrary. These academic fields have also been able to develop intrinsic abilities and behaviours that exclude them from others. It does not serve humanity to lock oneself in a single department when there is a world of novelty out there. Changing this paradigm, just like other paradigms, is affordable and easy because it only needs a split second for scales to fall and a new perspective to form.

Paradigms such as gradual change predicted from the gene-centred view of evolution by Dawkins, abrupt jumps punctuating periods of stasis by Steven Jay Gould and Niles Eldredge or even sociobiology by E. O. Wilson all then fit when we take the paradigm that OS presents. Organismal Selection states that there is an organism and its hellish universe. During times of increased populations, the organism increases in size through production. As in the case of biological species, they can be considered organizationally closed yet thermodynamically open, just like any other individual organism. In this sense E.O. Wilson's theory stands. At the same time, through increasing in size, primarily through growth and development and secondarily through production (i.e. reproduction), there is gradual, hidden change that cannot be appreciated by superficial physical examination. The karyotypes discovered in healthy and disease-bearing individuals and even in post-mortem examinations reveal that the gradualism is present. In this sense, the gene-centred view stands. Additionally, the initial division of organisms into those with organelles (eukaryotes) and those without (prokaryotes) is progressively being debunked (abrupt change) after a long period of stasis (i.e. the prevailing paradigm of the distinguishing factor between prokaryotes and eukaryotes). In various dimensions, these theories highlight gradualism at a genetic level and at a group level. An increase in size is possible through self-organization and by establishing stable hierarchical relationships, as organisms self-organize into states of criticality. Self-organization can be subtle at the genetic level, evident upon inspection using powerful scientific techniques or through simple observation and noting the behavioural group changes. As far as OS is considered, self-organization can also be conspicuous as when a star morphs into a black hole. These are all strategies for avoiding annihilation.

Then there is the asteroid attack that reshaped the face of the earth, eliminating more than 50% of known life forms. Such a harsh environment required the variability that is seen at the organismal level, whether as an individual or as a group. All of them are autonomous enough to assert their independence or dependence based on the imminent threat. This, through fossil record, explains the punctuated equilibria. These theories propagated by the eminent evolutionists might appear very different but when OS describes the various ways of looking at organisms and the robust map, we see that they are all on the same side albeit with different trajectories. An organism will seek to avoid annihilation since moments of stasis do not mean that the universe stops being brutal. It allows for the organism to tinker and build on its resilience during those times. Often, it might be subtle, not easily appreciated.

Studies have also demonstrated that we know only a meagre fraction of biodiversity on earth. Other forms of diversity could be conspicuous, as when communities chase away some people they might consider to be 'cursed beasts', just because a member had scarce hair or a different skin colour. Even then, in these situations, the universe can still hurl asteroids at the community or the 'cursed beast'. Since asteroids are also organisms, they also have their own problems. How organisms avoid all these forms of annihilation happens in different ways, enough to explain the gradual changes and the abrupt, punctuated shifts, either individually or in groups. Organismal Selection clarifies that the disparity between these theories might have been non-existent or small if ever it was present. They have been explaining narrow albeit accurate situations, thinking that it applies in the grand scheme. More than anything, it shows how paradigms can be so powerful as to set several scientists against each other, leading them apart enough for them not to see that they are on the same side. I myself, at the start of this book, thought that group selection could not be any superior but groups are everywhere and are organisms in their own right. Per Bak even showed how cataclysmic events match the production events that OS stresses. It brings us to the next aspect about paradigms, that is, correlation and causation.

Paradigms reflect why there are two schools of thought regarding correlation and causation. Correlation is often taken in bad taste compared to causality, when the difference is just that: taste. If there was an alien form that was able to view only the length of skirts and the flow of time, then by establishing the average length of skirts it can predict, with reasonable accuracy, the timeline when a group of human beings existed. It is a model that works very well but with correlation. It does not mean that time caused the shortening of skirts or vice versa. But it is a fairly accurate model. It is what Stephen Hawking laudably calls model-dependent realism. That should be the goal of seeking knowledge – finding a model that fits the data, or in short, accurate ones. Another might want to find a model that fits the rigors of causality. That is also encouraged. But the continuous argument that correlation is not causality should not be used to discourage correlation. In fact, the way we develop causative ideas often starts with correlation. When I come up with a theory such as OS, and how it amazingly fits the available data, there are bound to be differences out of firmly held conviction. It is expected. Nevertheless, efforts should be made to distinguish a model that speaks of correlation to avoid it being confused as causative.

Then again, the higher the leverage point, the higher the resistance against it. There will be causative diehards and in the advent of data mining, correlative prophets. Resistance is almost inevitable. For instance, space is made of just that, space! The mathematical models we create fail to do justice to the space between the planets in the solar system. Josh Worth has shown how a scaled version of the solar system is insanely large for our small minds to comprehend, let alone create models that scale to its exact size. An atom, despite its definition, which implies that it cannot be split, is largely space. If the hydrogen proton were the size of the sun we would need, on average, the diameter of 11 solar systems to capture to scale the average distance of the electron from the nucleus. It is difficult for our minds to comprehend such astronomical (atomic?) distances. Surely, an atom should just be as small as it is, as we have always known it to be. But scale up from the atom and the distance between buildings in a city is much closer than a proton is to the electron. The city is much more compact an organism than we might dismiss it. Our cellular connections are not as tightly linked as we like to imagine. Lovelock reminds us that the weak bonds that hold in our cellular membranes are ironically, what is holding us together. But such information seems bizarre. I found it bizarre. But once we're past the bizarre phase we can contemplate, at the very least, the possibility. In the same vein, my theory seems to contrast with that of Natural Selection, so I would expect resistance. In fact, I encourage it, in the interest of inquiry and ultimately, attaining accuracy.

My theory might be a form of threat to that of Natural Selection, even though I have elaborated how it explains the key concepts of Natural Selection. Even though Einstein's theory was able to explain the arc in Mercury's orbit, proving the more accurate theory of gravity, it does not dismiss Newtonian gravity. The paradigm of Natural Selection explains the various living phenomena as it has been understood since Darwin. Organismal Selection describes organisms differently, and I think, simply, to include the quark, the jaguar and the galaxy. In fact, OS helps me appreciate Natural Selection better by describing it differently. I hope it does the same to you. Much more than appreciation, the next leverage point is what I strive to live by.

The Power to Transcend Paradigms

There is the mathematician's habit of finding new proofs for what has previously been discovered. For instance, prime numbers have several proofs. The Pythagorean Theorem has very many proofs. It is an exercise that I admire, seeing that they unearth different proofs to test the validity of prime numbers or Pythagorean Theorem. The lesson is simple: not to fall in love with your paradigm. Not to fall in love (too much) with Natural Selection. Not to fall in love (too much) with Organismal Selection. I added the brackets because I can neither quantify nor elaborate what too much is for anyone any more than the other. The message, however, is that ideas tend to prevent thought. Having the ability to transcend them is the art of discovery. Even scientific discovery is tempered with this artistic drive. Super forecasters have even been noted to increase the accuracy of their prediction by having this same ability. By doing away with the fear of being wrong we get to explore the vast expanse of possibilities.

For instance, while exploring sexual selection, I discovered that I might have undersold the profound explanatory power of the handicap principle. My theory is very coarse-grained, compared to that described by Zahavi. I even felt that it has a more convincing reason for sexual selection compared to OS. In the process, I discovered that the same principle can also describes subtle ideas of mergers. If a lion sees a gazelle stotting, it might decide not to attack it. This is a reliable signal that many gazelles cannot fake. In the process, the role of the gazelle serves the lion. It preserves its energy by deciding to attack another gazelle that has not relayed the stotting signal. The role of the lion in interpreting the signal also serves the stotting gazelle. That is a stable hierarchical relationship, constituting a merger. As a result, both of them survive but in the herd of gazelles, there might be an unlucky one. It is such explorations that convinced me that a large bulk of evidence in support of the handicap principle might explain an unusual set of mergers among organisms. No biologist or evolutionist in his right might think that a lion and a gazelle might be an organism. Which form of imminent threat would you have to subject them to see if they resist? A savanna grassland fire perhaps. This organism comprising the lion and the gazelle might survive, but others might not. It goes to show that even in a herd, individuals assert their autonomy to prevent their annihilation by forming mergers with unlikely individuals. In short, sticking to a single explanation has the potential to blind one from alternative explanations.

Systems tend to be non-linear. Non-linear problems have more than one solution. For instance, biodiversity has been categorised into three levels– genetic, species and ecosystem. What OS states is that each of these categorisations is by fact true in the sense that each is an organism. From the example I have cited in the previous paragraph, there appears to be more diversity than just these three levels. Who would have thought that a merger between a lion and a gazelle would have existed, let alone constitute another string of biodiversity? Similarly, evolution ought to be explained by more than one theory. We had the gene-centred view, punctuated equilibrium, multi-level selection and I now introduce one that incorporates features of each, extending further into fields that were thought previously distinct from evolution. Had I not considered the possibility of particles as organisms, the idea of a common ancestor, an ancestor that we are unable to define, would still have remained the elusive focal point from which organisms had diverged. Nonetheless, we have not yet gotten a complete grasp of which organism was the first, only a description of how it behaved.

In the same vein, my theory describes both living and non-living organisms. I find that the theory remains consistent throughout my analysis. But I describe a non-linear phenomenon. If you feel that it does not explain the particulate world, well and good. We already have theories that adequately describe this world. However, the consistency applies to the biological organisms. This much should, at the very least, be considered by the specialists in biology, ecology and evolution.

As I finish, let us remember that evolution has been traced back to the very first organism, who until now has always been regarded as the common ancestor. The common ancestor is the single piece that links all the previous theories of evolution. But what if the ancestor was not able to 'reproduce' in the first instant? It is very likely. What if it failed several times before it was successful? It is very likely. What if 'reproduction' only happened when the organism sensed its pending size constraints? What if it is all *production* and not reproduction as we have always known it? What if, then, the sandpile and the gravid cat (let's call it Schrödinger) were similar in this sense? Wouldn't that make a particle an organism? Who, then, would have thought that the common ancestor, the first organism, had been a particle all along?

SOURCES OF ILLUSTRATIONS USED IN THE BOOK

1. Page 2: https://www.carleton.edu/departments/ENGL/Alice/ImageCcat.html

2. Page 14: https://www.chegg.com/homework-help/definitions/cell-structures-14

3. Page 21: https://www.physics.louisville.edu/cl-davis/phys111/notes/magn_field.html

4. Page 21: http://large.stanford.edu/courses/2017/ph241/udit2/

5. Page 39: Illustration by Kariuki James Mwangi

6. Page 47: https://laptrinhx.com/assumptions-of-linear-regression-837451595/

7. Page 50: https://biocyclopedia.com/index/algae/algae/chlorophyta.php

8. Page 55: https://www.birmingham.ac.uk/teachers/study-resources/stem/Physics/interference.aspx

9. Page 61: Illustration by Kariuki James Mwangi

10. Page 70: https://en.wikipedia.org/wiki/Quark

11. Page 72: https://www.pinterest.com/theronimlay/number-devil/

12. Page 74: https://www.physics.louisville.edu/cl-davis/phys111/notes/magn_field.html

13. Page 78: https://www.123freevectors.com/black-random-circles-dots-pattern-background-illustrator-100491/

14. Page 88: https://elifesciences.org/digests/67455/can-bacteria-predict-the-future

15. Page 89: https://www.researchgate.net/publication/283316463_Energy_Entropy_and_Complexity_Thermodynamic_and_information-theoretic_perspectives_on_ageing

16. Page 91: Illustration by Kariuki James Mwangi

17. Page 100: https://www.albertvieille.com/produit/bois-de-rose-perou/

18. Page 102: Source: Illustration by Kariuki James Mwangi

19. Page 108: https://professorsandcastle.com/page/5/

20. Page 116: https://www.bostonglobe.com/lifestyle/2016/11/06/petscolumn/HxzwsTZQWpCpfV4GL2kFiK/story.html

21. Page 120: https://www.martin-missfeldt.com/drawing-lines/drawing-mantis-shrimp

22. Page 121: https://pnghut.com/png/sa3Yy8A0T7/vampire-bat-illustration-black-and-white-transparent-png

23. Page 124: https://polyschemsurvival-guide.weebly.com/unit-7.html

24. Page 127: https://www.britannica.com/animal/eastern-coral-snake/images-videos#Images

25. Page 129: (a) https://www.niddk.nih.gov/news/media-library/16564 (b): https://immunesystem-sac.weebly.com/second-line-of-defence.html

26. Page 143: Illustration by Kariuki James Mwangi

27. Page 150: Illustration by Kariuki James Mwangi

28. Page 154: http://large.stanford.edu/courses/2017/ph241/udit2/

29. Page 156: Niklas KJ, Kutschera U. Kleiber's Law: How the Fire of Life ignited debate, fueled theory, and neglected plants as model organisms. Plant Signaling & Behavior.

2015 ;10(7):e1036216. DOI: 10.1080/15592324.2015.1036216.
PMID: 26156204; PMCID: PMC4622013.

30. Page 164: http://charles.vassallo.pagesper-
so-orange.fr/en/art/dimension.html

31. Page 170: Illustration by Kariuki James Mwangi

32. Page 175: Illustration by Kariuki James Mwangi

33. Page 189: https://www.li-sci.com/2022/06/polyploidie.html

34. Page 194: https://cen.acs.org/biological-chemistry/
origins-of-life/Iron-catalyze-metabolic-reactions-without/97/i18

35. Page 209: https://doi.org/10.1177/20597991221100427

36. Page 215: https://www.sciencedirect.com/
science/article/pii/B0122268652000444

37. Page 234: Journal of Mammalogy (Gene L & Barbara HF. Sexual
Behavior of the Gray Short-Tailed Opossum (Monodelphis
domestica), 1982) https://doi.org/10.2307/1380437
http://sibr.com/mammals/M001.html

38. Page 240: https://www.researchgate.net/figure/The-
CLAW-hypothesis-a-Gaian-feedback-mechanism-DMS-is-
produced-in-marine-ecosystem-and_fig2_258292437

39. Page 250: https://www.pinterest.com/
jessicablakney/peacock-images/

40. Page 267: Illustration by Kariuki James Mwangi

41. Page 268: https://donellameadows.org/archives/
leverage-points-places-to-intervene-in-a-system/

REFERENCES

Introduction

1. Mendel, G., 1965. Experiments in Plant Hybridisation. Harvard University Press.

2. Dawkins, R., 2016. The Selfish Gene. Oxford University Press.

3. Forbes, N., Mahon, B., 2014. Faraday, Maxwell, and the Electromagnetic Field: How Two Men Revolutionized Physics. Prometheus Books.

4. Lovelock, J., 2000. Gaia: The Practical Science of Planetary Medicine. Oxford University Press.

5. West, G., 2018. Scale: The Universal Laws of Life and Death in Organisms, Cities and Companies. Orion Publishing Group Limited.

Chapter 1

1. Pinker, S., 2010. The Language Instinct: How The Mind Creates Language. HarperCollins e-books.

2. Edwards, A.W.F., 2014. R.A. Fisher's gene-centred view of evolution and the Fundamental Theorem of Natural Selection. Biological Reviews 89, 135–147. https://doi.org/10.1111/brv.12047

3. Buerger, R., 2000. The Mathematical Theory of Selection, Recombination, and Mutation. John Wiley & Sons.

4. Traulsen, A., Nowak, M.A., 2006. Evolution of cooperation by multilevel selection. PNAS 103, 10952–10955. https://doi.org/10.1073/pnas.0602530103

5. Dawkins, R., 1996. The Blind Watchmaker: Why the Evidence of Evolution Reveals a Universe Without Design. Norton.

6. Wagner, A., 2014. Arrival of the Fittest: Solving Evolution's Greatest Puzzle. Penguin Publishing Group.

7. Darwin, C., 2009. On the Origin of Species. Cambridge University Press.

8. Caston, V., 2002. Aristotle on Consciousness. Mind 111, 751–815. https://doi.org/10.1093/mind/111.444.751

9. Hawking, S.W., 2011. The Illustrated Theory of Everything: The Origin and Fate of the Universe. Phoenix Books.

10. Watkins, J., 1995. Popper and Darwinism. Royal Institute of Philosophy Supplements 39, 191–206. https://doi.org/10.1017/S1358246100005506

11. Bak, P., 1996. How Nature Works: The Science of Self-organized Criticality. Springer.

Chapter 2

1. Teresi, D., n.d. Discover Interview: Lynn Margulis Says She's Not Controversial, She's Right. Discover Magazine. URL https://www.discovermagazine.com/the-sciences/discover-interview-lynn-margulis-says-shes-not-controversial-shes-right (accessed 8.13.20).

2. Hawking, S., Mlodinow, L., 2010. The Grand Design. Random House Publishing Group.

3. Strogatz, S., 2012. The Joy of x: A Guided Tour of Math, from One to Infinity. Houghton Mifflin Harcourt.

4. Smith, J.M., Holliday, R., 1979. Game theory and the evolution of behaviour. Proceedings of the Royal Society of London. Series B. Biological Sciences 205, 475–488. https://doi.org/10.1098/rspb.1979.0080

5. Harris, J.R., 2011. The Nurture Assumption: Why Children Turn Out the Way They Do. Simon and Schuster.

6. Krauss, L.M., 2012. A Universe from Nothing. Simon and Schuster.

7. Grosberg, R.K., Strathmann, R.R., 2007. The Evolution of Multicellularity: A Minor Major Transition? Annual Review of Ecology, Evolution, and Systematics 38, 621–654. https://doi.org/10.1146/annurev.ecolsys.36.102403.114735

8. Grosberg, R.K., Strathmann, R.R., 2007. The Evolution of Multicellularity: A Minor Major Transition? Annual Review of Ecology, Evolution, and Systematics 38, 621–654. https://doi.org/10.1146/annurev.ecolsys.36.102403.114735

9. Arnheim, R., 1974. Entropy and Art: An Essay on Disorder and Order. University of California Press.

10. Bell, J.S., 1964. On the Einstein Podolsky Rosen paradox. Physics Physique Fizika 1, 195–200. https://doi.org/10.1103/PhysicsPhysiqueFizika.1.195

11. Mermin, N.D., 1993. Hidden variables and the two theorems of John Bell. Rev. Mod. Phys. 65, 803–815. https://doi.org/10.1103/RevModPhys.65.803

12. Hoffman, D., 2020. Realism Is False | Edge.org [WWW Document]. URL https://www.edge.org/video/realism-is-false (accessed 8.14.20).

Chapter 3

1. Millien, V., 2011. Mammals Evolve Faster On Smaller Islands. Evolution 65, 1935–1944. https://doi.org/10.1111/j.1558-5646.2011.01268.x

2. Diamond, J., 2017. Guns, Germs, and Steel: The Fates of Human Societies. W. W. Norton & Company.

3. Hardin, G., 1993. Living within Limits: Ecology, Economics, and Population Taboos. Oxford University Press.

4. Kahneman, D., 2011. Thinking, Fast and Slow. Farrar, Straus and Giroux.

5. Popper, K.R., 2020. The Open Society and Its Enemies. Princeton University Press.

6. Gould, S.J., Vrba, E.S., 1982. Exaptation-A Missing Term in the Science of Form. Paleobiology 8, 4–15.

7. Raskin, I., 1992. Salicylate, A New Plant Hormone. Plant Physiol 99, 799–803.

8. Wagner, A., 2014. Arrival of the Fittest: Solving Evolution's Greatest Puzzle. Penguin Publishing Group.

9. Hillis, D., 2014. Edge.org. URL https://www.edge.org/response-detail/25435 (accessed 8.15.20).

10. Hillis, D., 2011. Edge.org. URL https://www.edge.org/response-detail/10071 (accessed 8.15.20).

11. WHO | World Health Organization, n.d.. WHO. URL http://www.who.int/neglected_diseases/vector_ecology/mosquito-borne-diseases/en/ (accessed 8.15.20).

12. Dawkins, R., 2016. The Selfish Gene. Oxford University Press.

13. Popper, K.R., 1979. Objective Knowledge: An Evolutionary Approach. Clarendon Press.

14. Clark, R.E., 2004. The classical origins of Pavlov's conditioning. Integr. psych. behav. 39, 279–294. https://doi.org/10.1007/BF02734167

15. Avery, J., 2012. Information Theory and Evolution. World Scientific.

16. Gribbin, J.R., 2004. Deep Simplicity: Bringing Order To Chaos And Complexity. Random House.

17. Turing, A.M., 1990. The chemical basis of morphogenesis. Bltn Mathcal Biology 52, 153–197. https://doi.org/10.1007/BF02459572

18. Murray, J.D., 2013. Mathematical Biology. Springer Science & Business Media.

19. Goodwin, B., 2001. How the Leopard Changed Its Spots: The Evolution of Complexity. Princeton University Press.

20. Green, J.B.A., Sharpe, J., 2015. Positional information and reaction-diffusion: two big ideas in developmental biology combine. Development 142, 1203–1211. https://doi.org/10.1242/dev.114991

21. Rhett, A., 2013. Is Light a Wave or a Particle?. WIRED. URL https://www.wired.com/2013/07/is-light-a-wave-or-a-particle/ (accessed 8.17.20).

22. Atkins, P., 2004. Galileo's Finger: The Ten Great Ideas of Science. Oxford University Press.

Chapter 4

1. Gould, S.J., Eldredge, N., 1977. Punctuated Equilibria: The Tempo and Mode of Evolution Reconsidered. Paleobiology 3, 115–151.

2. Pagel, M., 2014. Genetics: The neighbourly nature of evolution. Nature 514, 34–34. https://doi.org/10.1038/514034a

3. Gibson, B., Wilson, D.J., Feil, E., Eyre-Walker, A., 2018. The distribution of bacterial doubling times in the wild. Proc Biol Sci 285. https://doi.org/10.1098/rspb.2018.0789

4. Sumner, S., 2014. Edge.org. URL https://www.edge.org/response-detail/25533 (accessed 8.18.20).

5. Iourov, I.Y., Vorsanova, S.G., Yurov, Y.B., 2008. Chromosomal mosaicism goes global. Molecular Cytogenetics 1, 26. https://doi.org/10.1186/1755-8166-1-26

6. Youssoufian, H., Pyeritz, R.E., 2002. Mechanisms and consequences of somatic mosaicism in humans. Nat Rev Genet 3, 748–758. https://doi.org/10.1038/nrg906

7. Forsberg, L.A., Gisselsson, D., Dumanski, J.P., 2017. Mosaicism in health and disease — clones picking up speed. Nat Rev Genet 18, 128–142. https://doi.org/10.1038/nrg.2016.145

8. Somatic mosaicism in healthy human tissues, 2011. Trends in Genetics 27, 217–223. https://doi.org/10.1016/j.tig.2011.03.002

9. Margulis, L., Fester, R., 1991. Symbiosis as a Source of Evolutionary Innovation: Speciation and Morphogenesis. MIT Press.

10. Shively, J.M., Ball, F., Brown, D.H., Saunders, R.E., 1973. Functional Organelles in Prokaryotes: Polyhedral Inclusions (Carboxysomes) of Thiobacillus neapolitanus. Science 182, 584–586.

11. Gray, M.W., 2012. Mitochondrial Evolution. Cold Spring Harb Perspect Biol 4, a011403. https://doi.org/10.1101/cshperspect.a011403

12. Gibbs, S.P., 1981. The chloroplasts of some algal groups may have evolved from endosymbiotic eukaryotic algae. Annals of the New York Academy of Sciences 361, 193–208. https://doi.org/10.1111/j.1749-6632.1981.tb54365.x

13. Taylor, M., Mediannikov, O., Raoult, D., Greub, G., 2012. Endosymbiotic bacteria associated with nematodes, ticks and amoebae. FEMS Immunol Med Microbiol 64, 21–31. https://doi.org/10.1111/j.1574-695X.2011.00916.x

14. Margulis, L., Sagan, D., 2008. Acquiring Genomes: A Theory Of The Origin Of Species. Basic Books.

15. Pauli, W., 1994. Exclusion Principle and Quantum Mechanics, in: Writings on Physics and Philosophy. Springer, Berlin, Heidelberg, pp. 165–181.

16. Mandelbrot, B.B., 1983. The Fractal Geometry of Nature. Henry Holt and Company.

17. Darwin, C., 1871. The Descent of man. D. Appleton and Company.

18. Kimura, M., 1994. Population Genetics, Molecular Evolution, and the Neutral Theory: Selected Papers. University of Chicago Press.

19. Barton, N.H., Mallet, J., Clarke, B.C., Grant, P.R., 1996. Natural selection and random genetic drift as causes of evolution on islands. Philosophical Transactions of the Royal Society of London. Series B: Biological Sciences 351, 785–795. https://doi.org/10.1098/rstb.1996.0073

20. Kondepudi, D., Prigogine, I., 2014. Modern Thermodynamics: From Heat Engines to Dissipative Structures. John Wiley & Sons.

21. Tarnita, C., 2018. Ecology: Termite Patterning at Multiple Scales. Current Biology 28, R1394–R1396. https://doi.org/10.1016/j.cub.2018.10.058

Chapter 5

1. Bodenschatz, E., Pesch, W., Ahlers, G., 2000. Recent Developments in Rayleigh-Bénard Convection. Annual Review of Fluid Mechanics 32, 709–778. https://doi.org/10.1146/annurev.fluid.32.1.709

2. Prigogine, I., 1975. Dissipative structures, dynamics and entropy. International Journal of Quantum Chemistry 9, 443–456. https://doi.org/10.1002/qua.560090854

3. Ramirez, A., n.d. Interrogating and Shaping the World Through Science | Edge.org. URL https://www.edge.org/conversation/ainissa_ramirez-interrogating-and-shaping-the-world-through-science (accessed 8.21.20).

4. Yates, F.E., 2012. Self-Organizing Systems: The Emergence of Order. Springer Science & Business Media.

5. Holland, J.H., 2000. Emergence: From Chaos to Order. Oxford University Press.

6. Tenover, F.C., 2006. Mechanisms of Antimicrobial Resistance in Bacteria. The American Journal of Medicine 119, S3–S10. https://doi.org/10.1016/j.amjmed.2006.03.011

7. Lovelock, J., Lovelock, J.E., 2000. Gaia: The Practical Science of Planetary Medicine. Oxford University Press.

8. Meadows, D., 2008. Thinking in Systems: A Primer. Chelsea Green Publishing.

9. Thomas, C.M., Nielsen, K.M., 2005. Mechanisms of, and Barriers to, Horizontal Gene Transfer between Bacteria. Nat Rev Microbiol 3, 711–721. https://doi.org/10.1038/nrmicro1234

10. Topol, E., 2014. Edge.org. URL https://www.edge.org/response-detail/25363 (accessed 8.21.20).

11. Harris, J.R., 2010. No Two Alike: Human Nature and Human Individuality. W. W. Norton.

12. Harris, J.R., 2011. The Nurture Assumption: Why Children Turn Out the Way They Do. Simon and Schuster.

13. Capra, F., 1996. The Web of Life: A New Scientific Understanding of Living Systems. Anchor Books.

14. Cronin, T.W., Marshall, N.J., 1989. A retina with at least ten spectral types of photoreceptors in a mantis shrimp. Nature 339, 137–140. https://doi.org/10.1038/339137a0

15. Cronin, T.W., Caldwell, R.L., Marshall, J., 2001. Tunable colour vision in a mantis shrimp. Nature 411, 547–548. https://doi.org/10.1038/35079184

16. Obrist, M.K., 1995. Flexible bat echolocation: the influence of individual, habitat and conspecifics on sonar signal design. Behav Ecol Sociobiol 36, 207–219. https://doi.org/10.1007/BF00177798

17. Thomson, W., 1884. Steps towards Kinetic Theory of Matter. Science 4, 204–206.

18. Diamond, J.M., 1992. The Third Chimpanzee. HarperCollins.

19. Sherratt, T.N., 2002. The coevolution of warning signals. Proceedings of the Royal Society of London. Series B: Biological Sciences 269, 741–746. https://doi.org/10.1098/rspb.2001.1944

20. Charlesworth, B., 1994. The Genetics of Adaptation: Lessons from Mimicry. The American Naturalist 144, 839–847.

21. Pfennig, D.W., Mullen, S.P., 2010. Mimics without models: causes and consequences of allopatry in Batesian mimicry complexes. Proc Biol Sci 277, 2577–2585. https://doi.org/10.1098/rspb.2010.0586

22. Strogatz, S., 2020. Cori Bargmann on the Genetics of Transparent Worms, Supertasters and Cancer. Quanta Magazine. URL https://www.quantamagazine.org/cori-bargmann-on-the-genetics-of-transparent-worms-supertasters-and-cancer-20200317/ (accessed 8.23.20).

23. Randolph, G.J., Angeli, V., Swartz, M.A., 2005. Dendritic-cell trafficking to lymph nodes through lymphatic vessels. Nat Rev Immunol 5, 617–628. https://doi.org/10.1038/nri1670

24. Mayer, E.A., Knight, R., Mazmanian, S.K., Cryan, J.F., Tillisch, K., 2014. Gut Microbes and the Brain: Paradigm Shift in Neuroscience. J. Neurosci. 34, 15490–15496. https://doi.org/10.1523/JNEUROSCI.3299-14.2014

25. Babloyantz, A., 1977. Self-organization phenomena resulting from cell-cell contact. Journal of Theoretical Biology 68, 551–561. https://doi.org/10.1016/0022-5193(77)90105-9

Chapter 6

1. Poncela, J., Gómez-Gardeñes, J., Floría, L.M., Moreno, Y., 2007. Robustness of cooperation in the evolutionary prisoner\textquo-

tesingles dilemma on complex networks. New J. Phys. 9, 184–184. https://doi.org/10.1088/1367-2630/9/6/184

2. Koch, A.L., 1990. Diffusion The Crucial Process in Many Aspects of the Biology of Bacteria, in: Advances in Microbial Ecology, Advances in Microbial Ecology. Springer, Boston, MA, pp. 37–70.

3. Einstein, A., 2006. Relativity: The Special and the General Theory. Penguin Publishing Group.

4. Hawking, S., Mlodinow, L., 2011. The Grand Design. Bantam Books.

5. Buffenstein, R., 2008. Negligible senescence in the longest living rodent, the naked mole-rat: insights from a successfully aging species, J Comp Physiol B 178, 439–445. https://doi.org/10.1007/s00360-007-0237-5

6. Buffenstein, R., 2005. The Naked Mole-Rat: A New Long-Living Model for Human Aging Research. J Gerontol A Biol Sci Med Sci 60, 1369–1377. https://doi.org/10.1093/gerona/60.11.1369

7. Epelbaum, E., Krebs, H., Lähde, T.A., Lee, D., Meißner, U.-G., 2013. Dependence of the triple-alpha process on the fundamental constants of nature. Eur. Phys. J. A 49, 1–15. https://doi.org/10.1140/epja/i2013-13082-y

8. Guerrero, R., Margulis, L., Berlanga, M., 2013. Symbiogenesis: the holobiont as a unit of evolution. Int Microbiol 16, 133–143. https://doi.org/10.2436/20.1501.01.188

9. Li, K., Torres, C.E., Thomas, K., Rossi, L.F., Shen, C.-C., 2011. Slime mold inspired routing protocols for wireless sensor networks. Swarm Intell 5, 183–223. https://doi.org/10.1007/s11721-011-0063-y

10. Bonner, J.T., Slifkin, M.K., 1949. A Study of the Control of Differentiation: The Proportions of Stalk and Spore Cells in the Slime Mold Dictyostelium discoideum. American Journal of Botany 36, 727–734. https://doi.org/10.2307/2438226

11. Akaike, H., 1985. Prediction and Entropy, in: Selected Papers of Hirotugu Akaike, Springer Series in Statistics. Springer, New York, NY, pp. 387–410.

12. Lebowitz, J.L., 1993. Boltzmann's Entropy and Time's Arrow. Rutgers, The State University of New Jersey. https://doi.org/10.1063/1.881363

13. Cropper, W.H., 1986. Rudolf Clausius and the road to entropy. American Journal of Physics 54, 1068–1074. https://doi.org/10.1119/1.14740

14. Roszak, T., 1994. The Cult of Information: A Neo-Luddite Treatise on High-Tech, Artificial Intelligence, and the True Art of Thinking. University of California Press.

15. Diamond, J.M., 1998. Why Is Sex Fun?: The Evolution of Human Sexuality. Basic Books.

16. Keane, P.A., Seoighe, C., 2016. Intron Length Coevolution across Mammalian Genomes. Mol Biol Evol 33, 2682–2691. https://doi.org/10.1093/molbev/msw151

17. Souza, S.J. de, 2003. The Emergence of a Synthetic Theory of Intron Evolution. Genetica 118, 117–121. https://doi.org/10.1023/A:1024193323397

18. Tout, C.A., 2006. The triple-α process and the origin of the elements. Contemporary Physics 47, 145–155. https://doi.org/10.1080/00107510600715122

19. West, G.B., 2017. Scale: The Universal Laws of Growth, Innovation, Sustainability, and the Pace of Life in Organisms, Cities, Economies, and Companies. Penguin Press.

20. Kleiber, M., 1932. Body size and metabolism. Hilgardia 6, 315–353.

21. Koonin, E.V., 2010. The origin and early evolution of eukaryotes in the light of phylogenomics. Genome Biology 11, 209. https://doi.org/10.1186/gb-2010-11-5-209

22. Mandelbrot, B., 1967. How Long Is the Coast of Britain? Statistical Self-Similarity and Fractional Dimension. Science 156, 636–638. https://doi.org/10.1126/science.156.3775.636

23. Sancar, A., Lindsey-Boltz, L.A., Ünsal-Kaçmaz, K., Linn, S., 2004. Molecular Mechanisms of Mammalian DNA Repair and the DNA

Damage Checkpoints. Annual Review of Biochemistry 73, 39–85. https://doi.org/10.1146/annurev.biochem.73.011303.073723

24. Wallace, R., Wallace, D., Wallace, R.G., 2009. Eigen's paradox, in: Farming Human Pathogens. Springer, New York, NY, pp. 1–11.

25. Maturana, H.R., Varela, F.J., 2012. Autopoiesis and Cognition: The Realization of the Living. Springer Science & Business Media.

26. Varela, F.G., Maturana, H.R., Uribe, R., 1974. Autopoiesis: The organization of living systems, its characterization and a model. Biosystems 5, 187–196. https://doi.org/10.1016/0303-2647(74)90031-8

27. Eigen, M., Schuster, P., 2012. The Hypercycle: A Principle of Natural Self-Organization. Springer Science & Business Media.

28. Kauffman, S.A., Kauffman, M. of the S.F.I. and P. of B.S.A., 1993. The Origins of Order: Self-organization and Selection in Evolution. Oxford University Press.

29. Schrodinger, E., 2012. What is Life?: With Mind and Matter and Autobiographical Sketches. Cambridge University Press.

30. Lovelock, J.E., Margulis, L., 1974. Atmospheric homeostasis by and for the biosphere: the gaia hypothesis. Tellus 26, 2–10. https://doi.org/10.3402/tellusa.v26i1-2.9731

31. Krauss, L.M., 2012. A Universe from Nothing: Why There Is Something Rather Than Nothing. Simon and Schuster.

32. Shulman, M., Wilde, C.D., Köhler, G., 1978. A better cell line for making hybridomas secreting specific antibodies. Nature 276, 269–270. https://doi.org/10.1038/276269a0

33. Slattery, M.L., Boucher, K.M., Caan, B.J., Potter, J.D., Ma, K.-N., 1998. Eating Patterns and Risk of Colon Cancer. Am J Epidemiol 148, 4–16. https://doi.org/10.1093/aje/148.1.4-a

34. Pfeifer, G.P., Denissenko, M.F., Olivier, M., Tretyakova, N., Hecht, S.S., Hainaut, P., 2002. Tobacco smoke carcinogens, DNA damage and p53 mutations in smoking-associated cancers. Oncogene 21, 7435–7451. https://doi.org/10.1038/sj.onc.1205803

35. Peto's Paradox: evolution's prescription for cancer prevention, 2011. Trends in Ecology & Evolution 26, 175–182. https://doi.org/10.1016/j.tree.2011.01.002

36. Baish, J.W., Jain, R.K., 2000. Fractals and Cancer. Cancer Res 60, 3683–3688.

37. Kim, J., Dang, C.V., 2006. Cancer's Molecular Sweet Tooth and the Warburg Effect. Cancer Res 66, 8927–8930. https://doi.org/10.1158/0008-5472.CAN-06-1501

38. Feng, Y., Fuentes, D., Hawkins, A., Bass, J., Rylander, M.N., Elliott, A., Shetty, A., Stafford, R.J., Oden, J.T., 2009. Nanoshell-mediated laser surgery simulation for prostate cancer treatment. Engineering with Computers 25, 3–13. https://doi.org/10.1007/s00366-008-0109-y

39. Singhal, S., Nie, S., Wang, M.D., 2010. Nanotechnology Applications in Surgical Oncology. Annual Review of Medicine 61, 359–373. https://doi.org/10.1146/annurev.med.60.052907.094936

40. Masters, J.R., 2002. HeLa cells 50 years on: the good, the bad and the ugly. Nat Rev Cancer 2, 315–319. https://doi.org/10.1038/nrc775

41. Life expectancies - Office for National Statistics, n.d. URL https://www.ons.gov.uk/peoplepopulationandcommunity/birthsdeathsandmarriages/lifeexpectancies (accessed 8.27.20).

42. Bartlett, J.G., Chang, T.W., Gurwith, M., Gorbach, S.L., Onderdonk, A.B., 1978. Antibiotic-Associated Pseudomembranous Colitis Due to Toxin-Producing Clostridia. New England Journal of Medicine 298, 531–534. https://doi.org/10.1056/NEJM197803092981003

43. Barber, B.R., 2013. If Mayors Ruled the World: Dysfunctional Nations, Rising Cities. Yale University Press.

44. Moreira, D., López-García, P., 2009. Ten reasons to exclude viruses from the tree of life. Nat Rev Microbiol 7, 306–311. https://doi.org/10.1038/nrmicro2108

45. Pray, L.A., 2008. Discovery of DNA Double Helix: Watson and Crick | Learn Science at Scitable. URL https://www.nature.com/scitable/topicpage/discovery-of-dna-structure-and-function-watson-397/ (accessed 8.29.20).

46. Wilson, E.O., 1999. Consilience: The Unity of Knowledge. Vintage Books.

47. Luisi, P.L., Varela, F.J., 1989. Self-replicating micelles — A chemical version of a minimal autopoietic system. Origins Life Evol Biosphere 19, 633–643. https://doi.org/10.1007/BF01808123

48. Raoult, D., Audic, S., Robert, C., Abergel, C., Renesto, P., Ogata, H., Scola, B.L., Suzan, M., Claverie, J.-M., 2004. The 1.2-Megabase Genome Sequence of Mimivirus. Science 306, 1344–1350. https://doi.org/10.1126/science.1101485

49. Strogatz, S., 2020. Janna Levin on Seeing and Hearing Black Holes. Quanta Magazine. URL https://www.quantamagazine.org/janna-levin-on-seeing-and-hearing-black-holes-20200303/ (accessed 8.29.20).

50. Amelino-Camelia, G., 2000. Quantum theory's last challenge. Nature 408, 661–664. https://doi.org/10.1038/35047210

Chapter 7

1. Strogatz, S., 2012. The Joy of x: A Guided Tour of Math, from One to Infinity. Houghton Mifflin Harcourt.

2. Galilei, G., 1954. Dialogues Concerning Two New Sciences. Courier Corporation.

3. Taylor, R.S., 2010. Letter to the Editor. Paediatr Child Health 15, 258–258. https://doi.org/10.1093/pch/15.5.258

4. Lehingue, Y., Remontet, L., Munoz, F., Mamelle, N., 1998. Birth ponderal index and body mass index reference curves in a large population. American Journal of Human Biology 10, 327–340. https://doi.org/10.1002/(SICI)1520-6300(1998)10:3<327

5. Huerta, S., 2006. Galileo was Wrong: The Geometrical Design of Masonry Arches. Nexus Netw J 8, 25–52. https://doi.org/10.1007/s00004-006-0016-8

6. Bak, P., 1996. The Sandpile Paradigm, in: How Nature Works. Copernicus. pp. 49–64.

7. Simon, H.A., 1959. Theories of Decision-Making in Economics and Behavioral Science. The American Economic Review 49, 253–283.

8. Yanagimachi, R., 1981. Mechanisms of Fertilization in Mammals, in: Fertilization and Embryonic Development In Vitro. Springer, Boston, MA, pp. 81–182.

9. Staun-Ram, E., Shalev, E., 2005. Human trophoblast function during the implantation process. Reprod Biol Endocrinol 3, 1–12. https://doi.org/10.1186/1477-7827-3-56

10. Kearns, M., Lala, P.K., 1983. Life History of Decidual Cells: A Review*. American Journal of Reproductive Immunology 3, 78–82. https://doi.org/10.1111/j.1600-0897.1983.tb00219.x

11. Zhang, J., Dunk, C., Croy, A.B., Lye, S.J., 2016. To serve and to protect: the role of decidual innate immune cells on human pregnancy. Cell Tissue Res 363, 249–265. https://doi.org/10.1007/s00441-015-2315-4

12. Riemer, R.K., Heymann, M.A., 1998. Regulation of Uterine Smooth Muscle Function during Gestation. Pediatr Res 44, 615–627. https://doi.org/10.1203/00006450-199811000-00001

13. Li, Y., Reznichenko, M., Tribe, R.M., Hess, P.E., Taggart, M., Kim, H., DeGnore, J.P., Gangopadhyay, S., Morgan, K.G., 2009. Stretch Activates Human Myometrium via ERK, Caldesmon and Focal Adhesion Signaling. PLOS ONE 4, e7489. https://doi.org/10.1371/journal.pone.0007489

14. Loudon, J. a. Z., Sooranna, S.R., Bennett, P.R., Johnson, M.R., 2004. Mechanical stretch of human uterine smooth muscle cells increases IL-8 mRNA expression and peptide synthesis. Mol Hum Reprod 10, 895–899. https://doi.org/10.1093/molehr/gah112

15. Jacobeen, S., Pentz, J.T., Graba, E.C., Brandys, C.G., Ratcliff, W.C., Yunker, P.J., 2018. Cellular packing, mechanical stress and the evolution of multicellularity. Nature Phys 14, 286–290. https://doi.org/10.1038/s41567-017-0002-y

16. Caulin, A.F., Maley, C.C., 2011. Peto's Paradox: evolution's prescription for cancer prevention. Trends in Ecology & Evolution 26, 175–182. https://doi.org/10.1016/j.tree.2011.01.002

17. Nagy, J.D., Victor, E.M., Cropper, J.H., 2007. Why don't all whales have cancer? A novel hypothesis resolving Peto's paradox. Integr Comp Biol 47, 317–328. https://doi.org/10.1093/icb/icm062

18. Hill, R.A., Dunbar, R.I.M., 2003. Social network size in humans. Hum Nat 14, 53–72. https://doi.org/10.1007/s12110-003-1016-y

19. Gladwell, M., 2008. Outliers: The Story of Success. Little, Brown.

20. Anderson, M., 1984. The Evolution of Eusociality. Annual Review of Ecology and Systematics 15, 165–189. https://doi.org/10.1146/annurev.es.15.110184.001121

21. Alkema, L., Chou, D., Hogan, D., Zhang, S., Moller, A.-B., Gemmill, A., Fat, D.M., Boerma, T., Temmerman, M., Mathers, C., Say, L., 2016. Global, regional, and national levels and trends in maternal mortality between 1990 and 2015, with scenario-based projections to 2030: a systematic analysis by the UN Maternal Mortality Estimation Inter-Agency Group. The Lancet 387, 462–474. https://doi.org/10.1016/S0140-6736(15)00838-7

22. Nowak, M., Tarnita, C., Wilson, E., 2010. The evolution of eusociality. Nature 466, 1057–1062. https://doi.org/10.1038/nature09205

23. Brenner, B.M., Ballermann, B.J., Gunning, M.E., Zeidel, M.L., 1990. Diverse biological actions of atrial natriuretic peptide. Physiological Reviews 70, 665–699. https://doi.org/10.1152/physrev.1990.70.3.665

24. Adrian, E.D., 1914. The all-or-none principle in nerve. J Physiol 47, 460–474.

25. Rinella, M.E., 2015. Nonalcoholic Fatty Liver Disease: A Systematic Review. JAMA 313, 2263–2273. https://doi.org/10.1001/jama.2015.5370

26. Aggarwal, M., Khan, I.A., 2006. Hypertensive Crisis: Hypertensive Emergencies and Urgencies. Cardiology Clinics 24, 135–146. https://doi.org/10.1016/j.ccl.2005.09.002

27. Kiserud, T., Acharya, G., 2004. The fetal circulation. Prenatal Diagnosis 24, 1049–1059. https://doi.org/10.1002/pd.1062

28. Jauniaux, E., Jurkovic, D., Campbell, S., 1991. In vivo investigations of the anatomy and the physiology of early human placental circulations. Ultrasound in Obstetrics & Gynecology 1, 435–445. https://doi.org/10.1046/j.1469-0705.1991.01060435.x

29. Pang, W.W., Hartmann, P.E., 2007. Initiation of Human Lactation: Secretory Differentiation and Secretory Activation. J Mammary Gland Biol Neoplasia 12, 211–221. https://doi.org/10.1007/s10911-007-9054-4

30. Thornton, S., Davison, J.M., Baylis, P.H., 1988. Plasma oxytocin during third stage of labour: comparison of natural and active management. BMJ 297, 167–169. https://doi.org/10.1136/bmj.297.6642.167

31. Campbell, A., 2008. Attachment, aggression and affiliation: The role of oxytocin in female social behavior. Biological Psychology 77, 1–10. https://doi.org/10.1016/j.biopsycho.2007.09.001

32. Meyer-Reil, L.-A., 1977. Bacterial Growth Rates and Biomass Production, in: Microbial Ecology of a Brackish Water Environment, Ecological Studies. Springer, Berlin, Heidelberg, pp. 223–236.

33. Duina, A.A., Miller, M.E., Keeney, J.B., 2014. Budding Yeast for Budding Geneticists: A Primer on the Saccharomyces Cerevisiae Model System. Genetics 197, 33–48. https://doi.org/10.1534/genetics.114.163188

34. Conaway, C.H., 1958. Maintenance, Reproduction and Growth of the Least Shrew in Captivity. J Mammal 39, 507–512. https://doi.org/10.2307/1376787

35. Biggers, J.D., Curnow, R.N., Finn, C.A., McLAREN, A., 1963. REGULATION OF THE GESTATION PERIOD IN MICE. Reproduction 6, 125–138. https://doi.org/10.1530/jrf.0.0060125

36. Root Kustritz, M.V., 2006. Clinical management of pregnancy in cats. Theriogenology 66, 145–150. https://doi.org/10.1016/j.theriogenology.2006.03.018

37. Burris, M.J., Blunn, C.T., 1952. Some Factors Affecting Gestation Length and Birth Weight of Beef Cattle. J Anim Sci 11, 34–41. https://doi.org/10.2527/jas1952.11134x

38. Jafar, S.M., Chapman, A.B., Casida, L.E., 1950. Causes of Variation in Length of Gestation in Dairy Cattle. J Anim Sci 9, 593–601. https://doi.org/10.2527/jas1950.94593x

39. Bos, H., Mey, G.J.W. van der (Rijksuniversiteit U. (Netherlands) V.Z., 1980. Length of gestation periods of horses and ponies belonging to different breeds. Livestock Production Science (Netherlands).

40. Wittemyer, G., Barner Rasmussen, H., Douglas-Hamilton, I., 2007. Breeding phenology in relation to NDVI variability in free-ranging African elephant. Ecography 30, 42–50. https://doi.org/10.1111/j.0906-7590.2007.04900.x

41. Harder, J.D., Stonerook, M.J., Pondy, J., 1993. Gestation and placentation in two new world opossums: Didelphis virginiana and Monodelphis domestica. Journal of Experimental Zoology 266, 463–479. https://doi.org/10.1002/jez.1402660511

42. Wyatt, J., 2008. Anesthesia and Analgesia in Other Mammals 457–480. https://doi.org/10.1016/B978-012373898-1.50021-8

43. Baby Opossum n.d. URL https://www.wildlifecenter.org/baby-opossum (accessed 9.2.20).

44. Pollock, C., Arbona, N., 2018. Basic Information Sheet: Virginia Opossum. LafeberVet. URL https://lafeber.com/vet/basic-information-sheet-virginia-opossum/ (accessed 9.2.20).

45. Gestation period whales - Google Search [WWW Document], 2020. URL https://www.google.com/search?sxsrf=ALeKk03D3XsstjD-KyGbPdV1o7LdxAtOIpQ%3A1599051049970&ei=KZVPX4XVOt-DgUYj4r5AN&q=gestation+period+whales&oq=gestation+period+whales&gs_lcp=CgZwc3ktYWIQAzIECCMQJzICCAAyB-ggAEBYQHjIGCAAQFhAeMgYIABAWEB4yBggAEBYQHjIG-CAAQFhAeOgQIABBBHUJ_eEliV6xJgvvUSaABwA3gAgAG-2CIgBkhiSAQc0LTIuNy0ymAEAoAEBqgEHZ3dzLXdppesAB-AQ&sclient=psy-ab&ved=0ahUKEwjFjb6uwcrrAhVQcBQKHQ-j8C9IQ4dUDCA0&uact=5 (accessed 9.2.20).

46. Hildebrandt, T., Drews, B., Gaeth, A.P., Goeritz, F., Hermes, R., Schmitt, D., Gray, C., Rich, P., Streich, W.J., Short, R.V., Renfree,

M.B., 2007. Foetal age determination and development in elephants. Proc Biol Sci 274, 323–331. https://doi.org/10.1098/rspb.2006.3738

47. The Marine Mammal Center, n.d. URL https://www.marinemammalcenter.org/ (accessed 9.2.20).

48. Blue whale, 2010. Animals. URL https://www.nationalgeographic.com/animals/mammals/b/blue-whale/ (accessed 9.2.20).

49. Lee, P.C., Moss, C.J., 1986. Early Maternal Investment in Male and Female African Elephant Calves. Behavioral Ecology and Sociobiology 18, 353–361.

50. Diamond, J.M., 2005. Collapse: How Societies Choose to Fail or Succeed. Viking.

51. Schiermeier, Q., 2018. Gloomy 1970s predictions about Earth's fate still hold true. Nature. URL https://www.nature.com/articles/d41586-018-07117-2 (accessed 9.2.20).

52. Seymour, J.R., Simó, R., Ahmed, T., Stocker, R., 2010. Chemoattraction to Dimethylsulfoniopropionate Throughout the Marine Microbial Food Web. Science 329, 342–345. https://doi.org/10.1126/science.1188418

53. Capra, F., 1996. The Web of Life: A New Scientific Understanding of Living Systems. Anchor Books.

54. Sannino, A., Madaghiele, M., Carrozzo, M., Salvatore, L., Ambrosio, L., 2011. Nerve Tissue Engineering 435–453. https://doi.org/10.1016/B978-0-08-055294-1.00184-7

55. Strogatz, S., 2020. Cori Bargmann on the Genetics of Transparent Worms, Supertasters and Cancer. Quanta Magazine. URL https://www.quantamagazine.org/cori-bargmann-on-the-genetics-of-transparent-worms-supertasters-and-cancer-20200317/ (accessed 9.4.20).

56. Mackay, C.R., Marston, W.L., Dudler, L., 1990. Naive and memory T cells show distinct pathways of lymphocyte recirculation. J Exp Med 171, 801–817. https://doi.org/10.1084/jem.171.3.801

57. Darwin, C., 1876. Sexual Selection in Relation to Monkeys [WWW Document]. Nature. URL https://www.nature.com/articles/015018a0 (accessed 9.4.20).

58. Cuellar, O., 1977. Animal parthenogenesis. Science 197, 837–843. https://doi.org/10.1126/science.887925

59. Huigens, M.E., Luck, R.F., Klaassen, R.H.G., Maas, M.F.P.M., Timmermans, M.J.T.N., Stouthamer, R., 2000. Infectious parthenogenesis. Nature 405, 178–179. https://doi.org/10.1038/35012066

60. Bierzychudek, P., 1985. Patterns in plant parthenogenesis. Experientia 41, 1255–1264. https://doi.org/10.1007/BF01952068

61. Watts, P.C., Buley, K.R., Sanderson, S., Boardman, W., Ciofi, C., Gibson, R., 2006. Parthenogenesis in Komodo dragons. Nature 444, 1021–1022. https://doi.org/10.1038/4441021a

62. Morgan, M.T., 1994. Models of Sexual Selection in Hermaphrodites, Especially Plants. The American Naturalist 144, S100–S125. https://doi.org/10.1086/285655

63. Fisher, R.A., 1915. The evolution of sexual preference. Eugen Rev 7, 184–192.

64. Schwartz, S., 2015. Decoy switches frogs' mating call preference. Science News. URL https://www.sciencenews.org/article/decoy-switches-frogs-mating-call-preference (accessed 9.5.20).

65. Zahavi, A., Zahavi, A., 1999. The Handicap Principle: A Missing Piece of Darwin's Puzzle. Oxford University Press.

66. Borgia, G., 1985. Bower quality, number of decorations and mating success of male satin bowerbirds (Ptilonorhynchus violaceus): an experimental analysis. Animal Behaviour 33, 266–271. https://doi.org/10.1016/S0003-3472(85)80140-8

67. Dyer, S.J., 2009. The value of children in African countries – insights from studies on infertility. Journal of Psychosomatic Obstetrics & Gynecology. https://doi.org/10.1080/01674820701409959

68. Barry, K.L., Holwell, G.I., Herberstein, M.E., 2008. Female praying mantids use sexual cannibalism as a foraging strategy to increase fe-

cundity. Behav Ecol 19, 710–715. https://doi.org/10.1093/beheco/arm156

69. Kim, K.W., Roland, C., Horel, A., 2000. Functional Value of Matriphagy in the Spider Amaurobius ferox. Ethology 106, 729–742. https://doi.org/10.1046/j.1439-0310.2000.00585.x

70. Tucker, A., 2010. The Truth About Lions. Smithsonian Magazine. URL https://www.smithsonianmag.com/science-nature/the-truth-about-lions-11558237/ (accessed 9.6.20).

71. Mesce, D., 2018. Population Growth Concentrated Among the Poorest Communities – Population Reference Bureau. URL https://www.prb.org/population-growth-concentrated-among-the-poorest-communities/ (accessed 9.6.20).

72. Strassmann, B.I., 1996. The Evolution of Endometrial Cycles and Menstruation. The Quarterly Review of Biology 71, 181–220. https://doi.org/10.1086/419369

73. Jöchle, W., 1975. Current research in coitus-induced ovulation: a review. J Reprod Fertil Suppl 165–207.

74. Dawood, M.Y., 1981. Dysmenorrhoea and Prostaglandins: Pharmacological and Therapeutic Considerations. Drugs 22, 42–56. https://doi.org/10.2165/00003495-198122010-00003

75. Greendale, G.A., Lee, N.P., Arriola, E.R., 1999. The menopause. The Lancet 353, 571–580. https://doi.org/10.1016/S0140-6736(98)05352-5

76. Park, M., 2011. In giving life, women face deadly risks. CNN. URL https://www.cnn.com/2011/11/01/health/multiple-pregnancies-mother/index.html (accessed 9.6.20).

Chapter 8

1. Meadows, D., 2008. Thinking in Systems: A Primer. Chelsea Green Publishing.

2. Grove, A.S., 2010. Only the Paranoid Survive: How to Exploit the Crisis Points That Challenge Every Company. Crown.

3. Allison, A.C., 1954. Protection Afforded by Sickle-cell Trait Against Subtertian Malarial Infection. Br Med J 1, 290–294.

4. Williams, T.N., Mwangi, T.W., Wambua, S., Alexander, N.D., Kortok, M., Snow, R.W., Marsh, K., 2005. Sickle Cell Trait and the Risk of Plasmodium falciparum Malaria and Other Childhood Diseases. J Infect Dis 192, 178–186. https://doi.org/10.1086/430744

5. Gettes, L., 1992. Electrolyte abnormalities underlying lethal and ventricular arrhythmias. Circulation 85, I70–6.

6. Redman, C.W., Sargent, I.L., 2005. Latest Advances in Understanding Preeclampsia. Science 308, 1592–1594. https://doi.org/10.1126/science.1111726

7. George, R.H., Symonds, J.M., Dimock, F., Brown, J.D., Arabi, Y., Shinagawa, N., Keighley, M.R., Alexander-Williams, J., Burdon, D.W., 1978. Identification of Clostridium difficile as a cause of pseudomembranous colitis. Br Med J 1, 695.

8. Grajwer, L.A., Pildes, R.S., Horwitz, D.L., Rubenstein, A.H., 1977. Control of juvenile diabetes mellitus and its relationship to endogenous insulin secretion as measured by C-peptide immunoreactivity. The Journal of Pediatrics 90, 42–48. https://doi.org/10.1016/S0022-3476(77)80762-2

9. Treiber, T., Treiber, N., Meister, G., 2019. Regulation of microRNA biogenesis and its crosstalk with other cellular pathways. Nat Rev Mol Cell Biol 20, 5–20. https://doi.org/10.1038/s41580-018-0059-1

10. Ogawa, S., Lozach, J., Benner, C., Pascual, G., Tangirala, R.K., Westin, S., Hoffmann, A., Subramaniam, S., David, M., Rosenfeld, M.G., Glass, C.K., 2005. Molecular Determinants of Crosstalk between Nuclear Receptors and Toll-like Receptors. Cell 122, 707–721. https://doi.org/10.1016/j.cell.2005.06.029

11. Bateson, W., Mendel, G., 2013. Mendel's Principles of Heredity. Courier Corporation.

12. Sager, R., 1966. Mendelian and non-Mendelian heredity: a reappraisal. Proceedings of the Royal Society of London. Series B. Biological Sciences 164, 290–297. https://doi.org/10.1098/rspb.1966.0031

13. Rassoulzadegan, M., Grandjean, V., Gounon, P., Vincent, S., Gillot, I., Cuzin, F., 2006. RNA-mediated non-mendelian inheritance of an epigenetic change in the mouse. Nature 441, 469–474. https://doi.org/10.1038/nature04674

14. Harper, P.S., 1977. Mendelian inheritance or transmissible agent? The lesson Kuru and the Australia antigen. J Med Genet 14, 389–398.

15. Hernandez, D.G., Reed, X., Singleton, A.B., 2016. Genetics in Parkinson disease: Mendelian versus non-Mendelian inheritance. Journal of Neurochemistry 139, 59–74. https://doi.org/10.1111/jnc.13593

16. Wilson, E.O., 1984. The relation between caste ratios and division of labor in the ant genus Pheidole (Hymenoptera: Formicidae). Behav Ecol Sociobiol 16, 89–98. https://doi.org/10.1007/BF00293108

17. Johnston, A.B., Wilson, E.O., 1985. Correlates of Variation in the Major/Minor Ratio of the Ant, Pheidole dentata (Hymenoptera: Formicidae). Ann Entomol Soc Am 78, 8–11. https://doi.org/10.1093/aesa/78.1.8

18. Lodish, H., Berk, A., Zipursky, S.L., Matsudaira, P., Baltimore, D., Darnell, J., 2000. Noncovalent Bonds.

19. Bartee, L., Shriner, W., Creech, C., 2017. Chemical Bonds, in: Principles of Biology. Open Oregon Educational Resources.

20. Worth. If the Moon Were Only 1 Pixel - A tediously accurate map of the solar system. URL https://joshworth.com/dev/pixelspace/pixelspace_solarsystem.html (accessed 9.14.20).

21. Strogatz, S., 2012. The Joy of x: A Guided Tour of Math, from One to Infinity. Houghton Mifflin Harcourt.

22. Dawkins, R., 2014. Edge.org. URL https://edge.org/response-detail/25366 (accessed 9.14.20).

The First Organism hopes to offer a new and different perspective that would potentially change our understanding of the field of evolution. Using what can be regarded as the 4S's: Statistics, Symbiosis, Self-organization and Size, this summary explains a different dimension with which we can interpret evolutionary phenomena.

Darwin set the stage for a gradualist approach to evolution through Natural Selection. Dawkins introduced the world to a gene-centred view of evolution. Gould then punctuated his theory to explain the fossil record. E. O. Wilson later defended a social and biogeographical theory. Recently, Lamarck has been redeemed for his ideas of inheriting acquired traits. In the same spirit, *The First Organism* might all the more, leave you asking: Was there another organism other than the first one to compete with? Or to cooperate with? Does the first organism fall under all the taxonomic units? Or does it eliminate the idea of the phylogenetic tree? Does the idea of the first organism explain, at least in principle, the arrival of the fittest, the survival of the fittest, or both? Perhaps more questions are necessary than solutions towards the search for an alternative theory of evolution.

ISBN 978-9914-741-90-2

9 789991 474190 2